Electronic Engagement

A Guide for Public Sector Managers

Electronic Engagement

A Guide for Public Sector Managers

Dr Peter Chen

E PRESS

Published by ANU E Press
The Australian National University
Canberra ACT 0200, Australia
Email: anuepress@anu.edu.au
This title is also available online at: http://epress.anu.edu.au/engage_citation.html

National Library of Australia
Cataloguing-in-Publication entry

Chen, Peter.
E-engagement : a guide for public managers.

Bibliography
Includes index
ISBN 9781921313097 (pbk.)
ISBN 9781921313103 (online)

1. Business communication. 2. Digital communications. 3. Web sites - Design. 4. Information resources management.
I. Title.

658.45

All rights reserved. No part of this publication may be reproduced, stored in a retrieval system or transmitted in any form or by any means, electronic, mechanical, photocopying or otherwise, without the prior permission of the publisher.

Cover design by John Butcher

Funding for this monograph series has been provided by the Australia and New Zealand School of Government Research Program.

This edition © 2007 ANU E Press

John Wanna, *Series Editor*

Professor John Wanna is the Sir John Bunting Chair of Public Administration at the Research School of Social Sciences at The Australian National University. He is the director of research for the Australian and New Zealand School of Government (ANZSOG). He is also a joint appointment with the Department of Politics and Public Policy at Griffith University and a principal researcher with two research centres: the Governance and Public Policy Research Centre and the nationally-funded Key Centre in Ethics, Law, Justice and Governance at Griffith University. Professor Wanna has produced around 17 books including two national text books on policy and public management. He has produced a number of research-based studies on budgeting and financial management including: *Budgetary Management and Control* (1990); *Managing Public Expenditure* (2000), *From Accounting to Accountability* (2001) and, most recently, *Controlling Public Expenditure* (2003). He has just completed a study of state level leadership covering all the state and territory leaders — entitled *Yes Premier: Labor leadership in Australia's states and territories* — and has edited a book on Westminster Legacies in Asia and the Pacific — *Westminster Legacies: Democracy and responsible government in Asia and the Pacific*. He was a chief investigator in a major Australian Research Council funded study of the Future of Governance in Australia (1999-2001) involving Griffith and the ANU. His research interests include Australian and comparative politics, public expenditure and budgeting, and government-business relations. He also writes on Australian politics in newspapers such as *The Australian*, *Courier-Mail* and *The Canberra Times* and has been a regular state political commentator on ABC radio and TV.

Table of Contents

Foreword	xi
Preface	xiii
About the Author	xv
Acknowledgements	xvii
1. Introduction: An Information Age Democracy?	1
1.1. Who is this Guide For?	1
1.2. The Challenges of Engagement	3
1.2.1. An Expanding Policy Role for Public Sector Managers	4
1.3. The Information Society and its Implications	7
2. Definitions, Distinctions and Approaches to eEngagement	11
2.1. eDemocracy: A Conceptual Typology for Public Sector Managers	12
2.2. eEngagement as a Managerial Activity	14
2.3. Three Management Approaches	17
2.3.1. Active Listening	17
2.3.2. Cultivating	20
2.3.3. Steering	21
2.3.4. Relationship Between the Three Approaches	24
2.4. eEngagement and Electronic and Online Service Delivery	25
2.4.1. eGovernment Catalysts for eEngagement	27
2.4.2. Difficulties and Tensions	28
2.5. The Digital Divide: An Absolute Barrier?	29
2.5.1. Nature of the Divide	30
2.5.2. Implications of the Divide	32
2.5.3. Beyond the 'One Divide'	33
3. Designing the Right Approach	35
3.1. Key Decisions	36
3.1.1. What is the Issue(s)?	36
3.1.2. Who is the Audience(s)?	37
3.1.3. Consultation versus Collaboration	39
3.1.3.1. Implications of the Continuum	40
3.1.3.2. Reconceptualising Consultation and Collaboration	41

3.1.4. Setting Objectives	43
3.1.5. Degree of Interactivity	44
3.1.6. Choosing the Right Channel(s)	47
3.2. Concept Development Approach	48
3.3. Managing Identity Issues	50
3.3.1. Desirability of Identification	51
3.3.2. Technical Aspects of Identification	52
4. Implementation	55
4.1. Stakeholder Buy-in	55
4.2. Developing an Engagement Plan	55
4.3. Managing Technical Implementation	56
4.3.1. Determining the Software Feature Set	57
4.3.2. Who Governs? Technical, Administrative, or Political	58
4.3.3. Make or Buy?	60
4.3.3.1. Do we Need New Tools at All?	61
4.3.3.2. Purchase Point Considerations	62
4.3.3.3. Proprietary versus Open Source	63
4.3.4. Low Tech versus High Tech	65
4.4. Generating Compelling Content	67
4.4.1. Compelling Content versus Eyecandy	69
4.5. Promotion and Recruitment	70
4.5.1. Conventional Advertising and Promotional Approaches	71
4.5.2. The Power of Social Networking (and its Limitations)	71
4.6. Managing Risk	73
4.6.1. Security	74
4.6.2. Moderation	74
5. Concluding the Process	79
5.1. The Importance of Evaluation	79
5.1.1. Approaching Evaluation for eEngagement	80
5.1.2. Pitfalls to Avoid	80
5.1.3. What to Consider in Effective Assessment	81
5.2. Closeout Processes	82
5.2.1. Document Process and Outcomes	82

5.2.2. Feedback	83
5.2.3. Feedback Over Time	85
5.2.4. No Closeout: The Eternal Community	87
Further Reading	89
Appendix A. Policy Cycle Engagement Model	93
Appendix B. Catalogue of eEngagement Models	95

Foreword

In this monograph, Peter Chen has successfully performed a difficult balancing act by producing a coherent and comprehensive guide that is well grounded both conceptually and theoretically. Peter's task was made the more difficult by the fact that he is also writing about concepts that are highly contested, such as 'public value' or 'social capital'. Moreover, the policy and social landscape he traverses is continually evolving and shifting, driven by successive waves of emerging technologies and societal adaptations to technology-enabled communication.

Australians are enthusiastic adopters of mature technologies. It was once said, in the late 1980s and early 1990s, that Australia had more fax machines *per capita* than any developed industrialised country. Whatever the truth of such (possibly apocryphal) assertions, complex information and communication technologies (ICTs) now permeate every aspect of our daily lives. They are a crucial component of good public administration and policy delivery to the community.

The Australian public sector has embraced the promise of complex ICTs, albeit with perhaps greater reticence and less in the way of best practice than the private sector. But, as Chen argues, our public sectors are not necessarily the most adept embracers of ICTs. When ICTs are adopted, they may well not be utilised optimally or effectively, which may be explained partly by cultural and structural divides within organisations between technology guardians and end-users.

Chen examines the wide range of ICTs that might be employed to enable citizen engagement and participation in policy-making and program implementation. In addition to explaining the strengths and limitations of various ICTs, he explores the variety of circumstances in which they might be used and, importantly, alerts the reader to the opportunities they present or the pitfalls they entail.

The result is an engaging, provocative and thorough survey of available technologies and potential applications. This is a 'must read' for policy and program practitioners who are considering options for electronic engagement.

John Wanna
Sir John Bunting Chair of Public Administration
ANZSOG/ANU

Preface

The objective of this guide is to equip public sector managers to assess the value that new communications and computing technology may bring to their interactions with a range of potential stakeholders. It is written for managers who have an interest in expanding their approach to public engagement, rather than information technology professionals.

Over the last 20 years, advanced communication technologies, like computer networks and mobile telephones, have become pervasive throughout Western society. These technologies have not only revolutionised the delivery of public and private services, they have shaped consumers' expectations about service quality. These technologies can also play an important role in assisting public sector managers to consult, involve and engage members of the community in the development, implementation, management and evaluation of public policy.

This guide focuses on 'electronic engagement', which we might define as: 'the use of Information Communication Technologies by the public sector to improve, enhance and expand the engagement of the public in policy-making processes'.

This monograph does not advocate a specific methodology for electronic engagement. There is no single model that guarantees effective eEngagement. Instead, this guide emphasises the need to select, or develop a methodology that optimises four factors: *issue*, *audience*, *technology* and *timeframe*.

The incredible flexibility of new technologies provides the progressive public sector manager with a wide array of options for expanding their consultative and decision-making processes with key stakeholders. Public sector managers, however, need to consider a number of practical issues, including which approaches to electronic engagement are most appropriate to: (a) different management styles or roles; or (b) different points in the policy cycle. In so doing, managers might be well advised to catalogue the range of potential methodologies in a way that clearly sets out their advantages and limitations.

As a starting point for the development of an electronic engagement strategy, the guide discusses:

- motivations and reasons for expanding existing engagement strategies to incorporate new technologies;
- problems of definition and conceptualisation of these ideas, against the wider backdrop of the 'information society' and emerging 'electronically-facilitated democracies'; and
- management considerations, from initiation, development and implementation, to post-implementation review and assessment.

The guide includes a range of examples and references to other relevant manuals.

About the Author

Dr Peter Chen is a Research Associate with the National Centre for Australian Studies, part of the Faculty of Arts at Monash University.

Originally from Brisbane, Peter obtained his undergraduate qualifications in commercial management and public sector administration from Griffith University. While undertaking his doctorate in political science and public policy at The Australian National University, Peter worked on media regulation policy development associated with the emergent phenomenon of the internet.

Peter has professional experience in the public and private sectors. He has worked in policy development in the areas of telecommunications, the implications of demographic change for youth policy and law enforcement. In private consulting he worked on IT implementation and business process re-engineering as part of the Commonwealth's *Networking the Nation* project, focusing on systems procurement, website development and eCommerce for local governments.

In the field of electronic democracy, Peter served as the Inquiry Consultant for the Victorian Parliamentary *Inquiry into Electronic Democracy* in 2004-05 and is co-author of *Electronic Democracy? The Impact of New Communications Technology on Australian Democracy*, part of the Democratic Audit of Australia focussed audits series (2006).

Peter blogs at http://www.peterjohnchen.com/blog

Acknowledgements

The author would like to thank those policy practitioners who have contributed (directly and indirectly) to the development of this guide. Thanks too to ANZSOG for commissioning this monograph and to John Wanna and Jenny Keene who provided feedback on an early draft. I am grateful for the detailed comments provided on the finished manuscript by referees whose feedback improved the final version of the guide. Finally, special thanks goes to John Butcher and Bev Biglia at ANZSOG for their advice and editorial efforts.

1. Introduction: An Information Age Democracy?

Over the past 20 years, advances in information technology have had a significant impact on most societies. The scale of these impacts have been most profound in the developed world and have dramatically changed the way business is done in countries like Australia and New Zealand, which have something of a reputation as 'early adopters' of new technologies.

The ubiquity of communication and information technologies has significant implications for the ways in which the public sector conducts its business. The adoption of new technologies allows improvements in the delivery of public services, as well as the manner in which the work of the public sector is structured and undertaken.

Most governments in New Zealand and Australia have been active in developing new approaches to the delivery of public services using technologies like the internet and telephony. One area of growing public sector activity is the use of these technologies to improve communication with key stakeholder groups in order to engage them in public management decisions that affect their lives. The adoption of new technologies is, in part, a function of increasing accessibility and affordability. It also reflects a growing recognition of the *dynamic* and *interactive* potential of these technologies and their capacity to engage the public.

The adoption of new technologies is manifest in many ways, including:

- the substitution of old methods of communication with new ones;
- the development of new channels of communication with existing stakeholders;
- the ability to access new stakeholder groups and draw them into the policy development and implementation process;
- active participation in decision-making by the community; and
- new forms of policy administration and implementation using collaborative technologies.

Given the possibilities that new communications and computing technologies provide to public sector managers, this manual aims to introduce and discuss the concept of 'electronic engagement' (eEngagement), namely, the use of new technologies in a range of consultative and deliberative processes which enhance public participation in shaping policy outcomes.

1.1. Who is this Guide For?

This guide has been written for the public sector manager who has an interest in the ways in which information and communication technologies may be used to *democratise* decision-making and policy implementation. It is specifically written for *policy* professionals (rather than *information technology* professionals) who have a modest understanding of modern information technology.

Four types of public sector manager might want to use this guide:

- the manager facing a 'challenge' and possibly looking for alternative ideas or approaches to new or existing problems (including those specifically concerned with the challenges of consultation such as poor outcomes obtained in previous engagement processes);
- the manager with experience in online community engagement, which may have delivered sub-optimal outcomes and who is looking to review and refocus their efforts;
- managers with carriage of consultative processes looking to add new techniques to their repertoire; or
- public sector managers concerned with the effective implementation and evaluation of initiatives in the area of electronic democracy.

Unfortunately, this guide cannot provide one single implementation path for eEngagement, because:

- the range of activities that can fall under this category is extremely broad. Although some specific models have had significant attention to date, many other online engagement approaches are still at a formative and *experimental* stage of development;
- the speed of technological change is outstripping the capacity of policy researchers to theorise its application. It may be *years* before the full implications of the introduction of new media are realised; and
- the mix of *issue* , *methodology* , *audience* and *technology* often makes each implementation unique. Often 'tried and true' models translate poorly into different environments for a range of reasons, including cultural differences, excessive expectations and different institutional approaches to policy development.

Despite the virtual explosion of technological innovation eEngagement is not a mature field. Understanding the range of possibilities for these technologies will require the sharing of experiences, coupled with ongoing experimentation, evaluation and documentation. This is currently being undertaken by practitioners using online tools to share experiences and information and through institutional responses by governments setting up practice areas.

1.2. The Challenges of Engagement

The public sector is a set of institutions continually subject to reform pressures and new challenges. While the classic bureaucratic model of public administration stressed standardisation, rule-based management and stability, wider pressures on the public sector over the last 50 years have encouraged increased flexibility in program implementation and *inclusiveness* in program design.[1]

These changes tend to be driven by concerns about effectiveness and significant constraints on the resources made available to public sector managers. Throughout the public sector, it is increasingly recognised that the achievement of successful program outcomes requires:

- the incorporation of local actors and stakeholders in public programs;
- co-ordination of activities across organisations, jurisdictions and between public and private sectors; and
- the development and implementation of policy based on strong evidentiary justifications.

At the same time, however, public confidence in the role and ability, of government has been declining over recent years (see Exhibit 1). The public continues to have high expectations of the services provided by the public sector, but is increasingly sceptical of government's abilities and cynical regarding the motivations of political leaders.[2] This level of disengagement makes devolution difficult and undermines the recognition of positive public programs.

> **Exhibit 1: Declining Trust in Government — New Zealand and Australia**
>
> ... in 1985, 8.6% of New Zealanders had 'a great deal' of confidence in the government. By 1998 that figure had fallen to 2.5%. The number of people who were 'not at all' confident in the good intentions of their government doubled from 11.1% in 1985 to 21.8% in 1998.
> Barnes and Gill, 2000, *Declining Government Performance?*
>
> *Why Citizens Don't Trust Government*
>
> ... many public institutions lost public confidence in the period 1983-95, including the legal system 61% to 35%, the press 29% to 16%, the public service 47% to 38% and the Federal government 55% to 26%.
> Cox, 2003, *Social In/Equality*

---

[1] Thomas, John, 1995, *Public Participation in Public Decisions: New Skills and Strategies for Public Managers*, Jossey-Bass Publishers, San Francisco.
[2] Žižek, Slavoj, 1989, *Sublime Object of Ideology*, Verso, London.

1.2.1. An Expanding Policy Role for Public Sector Managers

Public scepticism about the role of the state has contributed to the emergence of managerial norms that emphasise inclusive and devolved policy development. In turn, this has required public sector managers to acquire new skills and capabilities.

While the classic bureaucratic model emphasised and rewarded strict technical expertise, the modern public sector manager is expected to have a range of 'soft' skills around coalition formation and stakeholder management. Some of the representative functions (consultation, negotiation, coalition building and other political skills) once attributed to political leaders have been delegated (appropriately or not) to public sector managers.

This approach is both consistent and broadly aligned with an empahasis on notions of *public value* in public sector management. Mark Moore, in his 1995 book *Creating Public Value* (see Exhibit 2), remarks on what he sees as a trend towards executive government expecting public sector managers to be *responsive* to the public's interests and concerns. Needless to say, this creates new responsibilities and accountabilities for managers.

> **Exhibit 2: Engagement as the Creation of Public Value**
>
> In his 1995 Book, *Creating Public Value*, Mark Moore outlines the role of the public manager as an agent in creating 'value' in the public sector, in the same way that a private-sector manager is tasked with creating private (or 'shareholder') value. Public managers, according to Moore, need to focus their attention on a bridging role between political leaders (who are 'authorisers' of manager's plans) and stakeholder groups.
>
> Thus, as in the private sector, alignment needs to be made between the satisfaction of those who consume the public services being provided (customers and clients) and those to whom the manager is directly responsible (political leadership as a 'board of directors'). This requires that public managers recognise themselves as an having an important role in 'strengthening the policies that are sold to their authorisers'. Moore, therefore, observes that the public manager is critical in shaping the public 'narrative' around their programs of action.
>
> While this can simply be seen as call for better program development in the public sector, Moore's notion of public managers as value creators is broader: First, he recognises that the analogy between public and private value creation is somewhat false and that public value creation is not the result of a free choice by clients, but suffers from coercion (either at the individual level of program consumption, or collectively

through taxation). The public manager is constrained in their ability to rely on 'marketplace' models of accountability.

Secondly, Moore is very critical of 'defects' in the political system as a guide to action for public managers. Citing 'corrupting' elements like short-termism, irrational decision-making and risk avoidance, Moore identifies a basic problem in classical conceptions of bureaucratic neutrality (*vis-à-vis* the criticisms of Max Weber as to the de-humanising nature of classic hierarchical bureaucracy).

Overall, therefore, the notion of public value creation recognises that there is a strong moral imperative placed on the public manager that results from their reliance on coercion for their provision of the resources at their disposal, matched with a need to balance the provision of immediate services to client groups against societal benefits (such as the aggregate benefit of their action, the effective choice of ends and political accountability).

In terms of engagement activities provide opportunities to leverage public value by:

- directly creating a means by which the public manager can develop their programmatic policy proposals through stakeholder input (at the design, development, reporting and post-implementation review stages);
- expanding the range of stakeholder groups that can form part of the narrative creation process (recognising and incorporating groups beyond direct authorisers and clients, but who may receive public value indirectly);
- accessing a constituency around the policy area directly to mobilise action where political failures or corruption becomes apparent and/or act entrepreneurially to strengthen policy proposals for authorisation; and
- developing a long-term view to ensure that public institutions and not simply immediate programs, are strengthened and sustainable (inter-generational value creation).

Moore, Mark. 1995. *Creating Public Value: Strategic Management in Government*. Cambridge: Harvard University Press.

An overarching focus on a *strategic planning* approach has been matched in recent years with a renewed concern for local community development and empowerment as a critical element in program implementation. Combined, these trends lead to greater:

- emphasis on public servants acting *entrepreneurially* to create public value through innovation and responsiveness across all levels of government;
- social inclusion and capacity building at the local level to develop various forms of *social capital* that sustains communities through capacity building and the development of community resilience (see Exhibit 3); and
- renewed interest in inter-organisational *collaboration and partnerships* between government, private-sector and community groups and across organisational divisions within government.

Exhibit 3: eEngagement and Social Capital – Chicken and Egg

Social Capital represents the idea that there is value in community. That is, that the social ties that bind communities and individuals together produce a range of positive benefits (psychological as well as material) that make vibrant communities more healthy and resistant to change than those with low levels of interpersonal contact, trust and shared norms and concerns.

In the public policy context, the recognition of the value of community ties is not new. Post-war policy failures associated with the US 'great society' projects of Lyndon B. Johnson identified a lack of planning around the social supports that brace policy delivery and the 1970s saw renewed interest in 'communities' as loci for policy development and implementation. Over the last decade and a half, the notion of social capital has renewed public-sector interest in the role governments can have in developing and strengthening communities. It is now common for public servants to talk of the role of governments in 'investing' in social capital, or the implications of different levels of social capital on program delivery and program effectiveness.

Social capital has clear implications for the eEngagement practitioner:[3] First, public engagement is fundamentally tied to social capital. This can be through the role of community-based engagement activities in developing new or renewed ties within the community (a 'bringing together' process), assisting in the development of shared understandings and values, or though the aggregation and exchange of resources held within the community. In this way, the consultative process of co-decision-making can move to one of co-production, where the social capital developed during consultation can be tasked towards partial or

---

[3] van den Hoof, Bart, de Ridder, Jan and Aukema, Eline. 2004. 'Exploring the Eagerness to Share Knowledge: The Role of Social Capital and ICT in Knowledge Sharing'. *Social Capital and Information Technology*. Huysman and Wulf (eds), MIT Press, Cambridge.

complete program delivery. The benefits that this can bring in terms of community 'ownership' of the resultant policy can be considerable.

Second, public engagement can be dependent on levels of social capital in the community: where social ties are weak and trust is low, public engagement processes need to invest considerably more time in establishing shared understanding of the 'rules of the game' (consultation and participation processes), the basic nature of the issues under consideration and earning of trust.

Overall, social capital is commonly associated with social networks and therefore has a strong relationship to networking technologies. In recent years the term 'social media' has come to describe a range of technologies (such as social networking websites, blogs, wikis) where users have a key role in the creation of content and therefore the value associated with the service. These services represent a good example of social capital: while the underlying technology (such as the website) serves as a facilitator for collective action and benefit, the participation of members and the relationship between members serves as the primary generator of value. For a good guide to social media, see: Cook, Trevor and Hopkins, Lee. 2006. *Social Media or, 'How I learned to stop worrying and love communication'*, <http://trevorcook.typepad.com/weblog/files/CookHopkins-SocialMediaWhitePaper.pdf>

1.3. The Information Society and its Implications

As we move into an era characterised by an abundance of information and communications channels, the growth of new information technologies has challenged many traditional assumptions about the relationship between government and citizens. The ubiquity of information technology in developed societies acts to reduce the public sector's significant control over on policy-related information and the interactive nature of these technologies empowers citizens to form coalitions, mobilise opinion and engage with decision-makers.

This may lead to the emergence of 'electronically-facilitated democracies': political jurisdictions where the political process is conducted through a variety of electronic systems. These developments also have a number of implications. First, public sector managers need to understand that the community has the capacity to be far better educated and informed than in any other period of time. This move towards an *information society* means that:

- members of the public are less willing to accept government decisions based on *appeals to authority* alone, transparency of decision-making is needed;

- they have greater means to challenge the analysis and decision-making of government through their own expertise and information — *participation is often demanded*; and
- *communities of interest* (stakeholder groups, interest groups) need not be geographically-defined, can arise spontaneously and can play active roles in facilitating (or preventing) implementation.

> **Exhibit 4: The Information Society – Definition**
>
> A society characterised by a high level of information intensity in the everyday life of most citizens, in most organisations and workplaces; by the use of common or compatible technology for a wide range of personal, social, educational and business activities; and by the ability to transmit and receive digital data rapidly between places irrespective of distance.
>
> IBM Community Development Foundation, 1997

Second, it must be recognised that the technological environment in which government operates is more complex and diverse. This can be seen in:

- a new emphasis on *multi-channel service delivery:* single-channel communications strategies are increasingly ineffective in reaching broad target audiences. Members of the public increasingly expect a flexible interface with government and they expect to be offered a choice of technologies through which to access services. This is particularly true of younger people who have grown up with internet technologies and mobile telephones;
- the *low cost* of electronic communication is significant. It increases the *speed* of information distribution, stimulates 'virtual' interest groups and provides an array of channels by which the public can communicate with elected representatives and public servants. The advent of email, in particular, has seen a growth in the correspondence received and generated by government. This is also manifest in 'information overload' and issues associated with the correct storage, presentation and indexing of information;[4] and
- new technologies *break down boundaries* between organisations and jurisdictions. Public interest groups are more engaged with 'sister' organisations around the world – sharing information and strategies, while public servants are using these technologies to increase policy-learning and professional networking.

While some governments have developed specific 'electronic democracy' policy agendas (e.g. Queensland), it is important to recognise that the development of

---

[4] Often referred to as the ease of 'discoverability' of information held in publicly-accessible information systems, or in corporate archives.

Australia and New Zealand as electronically-facilitated democracies is both *deliberate* and *organic* in character: governments can elect to use Information Communication Technologies (ICTs) instrumentally for consensus-building and policy development, but will also be subject to a range of demands from new groups in the community who have used these tools to mobilise politically.

The implications for effective and orderly public administration in information societies will be profound, requiring:

- public sector organisations that are responsive to their changing technological and social environment. This requires appropriate structures, skills and resources within these organisations;
- appropriate policy frameworks to foster and support the active role of the public in policy development and implementation;
- appropriately developed, managed and targeted electronic democracy initiatives to address issues of public disenchantment with government;
- learning organisations that recognise the highly fluid and largely unknown nature of the relationship between new technologies and policy processes; and
- clear recognition of the value of traditional forms of community consultation and engagement. New methods are introduced within the context of parliamentary democracy and many members of our community will continue to rely on 'conventional' media and participatory forms.

Being responsive to the communities' expectation of government communications will require an awareness of technological developments and community norms and expectations.

Exhibit 5: eDemocracy as an 'Evolving Concept'

'I think we often speak as if there is a completed project called "democracy" and there is another completed project called "the internet" and we ask "what will this thing called the internet do to this thing called democracy?". Both of these are in a state of evolution. We haven't got a completed democracy; we haven't got a completed internet. Both are up for grabs. So the question we need to ask is whether the internet is likely to reinforce traditional ways of doing politics, which has tended to be rather remote from the public. Or whether the internet, as an interactive medium, can enable the public to get into a more collaborative and conversational style of politics which makes it more meaningful to them.'

Professor Stephen Coleman, Oxford Internet Institute, 2004

2. Definitions, Distinctions and Approaches to eEngagement

When developing a management approach for *eEngagement*, one of the most common barriers faced by public sector managers in New Zealand and Australia is the wide array of competing, contested and conflicting definitions employed to describe it.

Even an increasingly common term like 'electronic democracy' evokes an array of responses, from highly specific definitions (such as voting over the internet) to nebulous concepts (an information environment which is open, participative and free to access). These terms can be loaded and be a vehicle for a variety of implicit assumptions and norms, particularly around issues of *direct democracy*.

> Exhibit 6: Direct Democracy – Definition
>
> A form of democratic government whereby citizens have the right to participate in decision-making through referenda on legislative initiatives. Direct democracy can exist in parallel to representative democracy, for example, where ballot initiatives allow citizens to vote on legislative initiatives, or replace representative democracy. In practice, direct democracy is limited by the complexity of modern policy making and the capacity for citizens to deliberate issues in a timely and expedient manner.
> Scrutiny of Acts and Regulations Committee, Parliament of Victoria, 2005

This emerging area of practice and study has generated a range of competing terms because the technology and its impact on political processes is so new. It may be many years, if ever, before scholarship and practice moves towards agreement on terminology. In addition, the complex and often ill defined nature of policy-making processes, combined with the highly dynamic nature of information technology, work against the establishment of a clear, unambiguous definition for *eEngagement*.

While this proliferation of terminology is confusing and sometimes only reflects the predilection of individual authors, some terms are carefully chosen and have distinct meanings based within a specific area of literature or practice. Readers need to take care when a term is deliberately employed because it may have a very specific meaning. A good example would be the differing use of the terms 'eDemocracy' and 'eGovernance'. The former commonly refers to a broader notion of equal participation throughout the political *system*,[1] while the latter can refer to an organisational or inter-organisational focus.[2]

Similarly, some authors use different terms in a nested, or typological, manner. Two examples would include:

- the use of 'eGovernment' to refer to the overarching application of information and communications technologies by government and 'eDemocracy' as those uses in government with a specific political focus; versus
- the use of 'eGovernment' to specifically refer to electronic and online service delivery and 'eDemocracy / eGovernance' to refer to policy-making processes utilising new technology.

Exhibit 7: The Confusing Terminology of eEngagement

Each of the following prefixes and suffixes has been used at one time or another to describe this area of practice (the list is not exclusive)

Prefix	*Suffix*
Electronic (e-)	Government
Online	Democracy
Digital	Governance
i- (as in information)	Engagement
Cyber	Commons
Virtual	Participation
Tele	Agora
Mobile (m-)	Rule Making

2.1. eDemocracy: A Conceptual Typology for Public Sector Managers

While there is value in separating the 'political' and 'technical' elements of public management, the investment in public sector infrastructure, electronic democracy initiatives and electronic service delivery are at once separate and complementary, activities.

---

[1] Which may include activities outside of the scope of government intervention or control, such as the creation of democratic 'alternative media' (community media), electronic activities and protest and democratic actions aimed at non-government institutions (other nations, corporate actors, etc.).

[2] 'Governance' is itself used with various meanings by different scholars and practitioners. From the perspective of an organisational theorist it often refers to the regulation of an organisation as a system with internal and external feedback and information collection mechanisms ('cybernetics'). From a socio-political perspective, it refers to networks of interdependent organisations that engage in complex bargaining relationships.

This separation results from a number of factors, trends and contradictions:

- there is often an explicit desire on the part of democratic theorists to separate service functions from democratic functions, due to a conceptual and philosophical delineation between notions of inherent political rights and the reciprocal and/or conditional relationships commonly implied in service provision;
- democratic participation has an emphasis on universalism (such as equal participation for all), whereas in developed nations there is an increasing emphasis on *selective* service delivery;
- there is often a managerial desire to maintain a separation of policy development from service functions, either due to the logic of purchaser-provider splits, or to separate payment functions from policy access;
- much of the overarching information technology infrastructure (the *technological* level) associated with electronic and online service delivery is of equal value in facilitating electronic participation and democracy: for example, encryption standards can be employed for *eProcurement* or for *online voting* (the *application* level); and
- the development and implementation of electronic and online service delivery systems is commonly undertaken by business process or customer service units, rather than policy development units.

A more useful way of conceptualising the relationship between the development of an electronically-facilitated democracy and the role of public sector managers as Moore's responsive entrepreneurs is presented in Figure 1. This figure associates different types of engagement activities with different management roles or 'approaches' to project implementation, based on two axes of classification:

- the *Nature of the Programmatic* approach: representing the expected role of government in programs which result from the engagement process (the degree to which project outcomes will be 'top down' or 'bottom up'); and
- the *Specificity of Outcome (Intention):* representing the degree to which eventual outcomes will be highly focused (with simple / singular performance criteria) or more diffuse in their objectives (resulting in more complex / perceptual performance reporting).

Figure 1: Conceptualising the Scope of eDemocracy

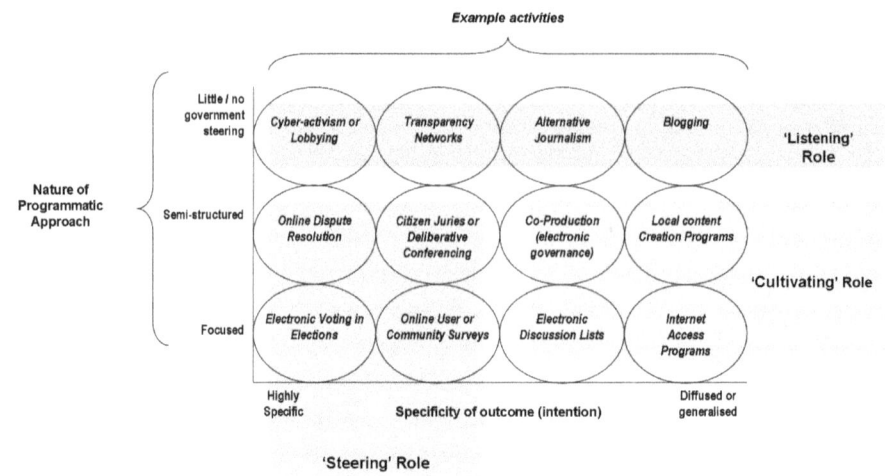

2.2. eEngagement as a Managerial Activity

Figure 1 shows how electronic democracy activities require different managerial approaches, depending on (a) the sphere in which primary activity occurs (state-centric versus societal) and (b) the objectives of the programmatic response of government. While all of the activities indicated in this figure have fundamental democratic outcomes and objectives, the role of policy-development units (and staff) in some of these areas is limited.

While activities like public access terminal placement programs provide democratic outcomes, the relationship between these programs and policy development activities is generally one-way. Public sector managers wanting to open up the policy-making process to public participation should clearly distinguish between the broad area of eDemocracy and particular applications of electronic engagement such as service kiosks.

In the context of this guide, 'Electronic Engagement' (eEngagement) is defined as:

> *The use of Information Communication Technologies (ICTs) by the public sector to improve, enhance and expand the engagement of the public in policy-making processes.*

This definition is at once broad and narrow in its scope. It is broad in that it:

- does not specifically relate to any particular *methodology* of engagement, such as *direct decision-making* or *online consultation*. These are methods that fall within its scope;

- focuses on the public sector (the bureaucracy) and its wide array of activities, needs and stakeholder groups; and
- includes an array of technologies, not simply the internet.

It is narrow in that it:

- does not include electoral processes and political campaigning (see Cornfield's 2004 *Politics Moves Online* or Browning's 2002 *Electronic Democracy)*;
- excludes areas of public sector activity related to technological access (see Servon's 2002 *Bridging the Digital Divide)* or the development of an 'information society' (see Norris's 2001 *Digital Divide).*

Exhibit 8: ICTs Defined

'Information and communications technologies (ICTs) is a term which is currently used to denote a wide range of services, applications and technologies, using various types of equipment and software, often running over telecom networks.

'ICTs include well known telecom services such as telephone, mobile telephone and fax. Telecom services used together with computer hardware and software form the basis for a range of other services, including email, the transfer of files from one computer to another and, in particular, the Internet, which potentially allows all computers to be connected, thereby giving access to sources of knowledge and information stored on computers worldwide.

'Applications include videoconferencing, teleworking, distance learning, management information systems, stock taking; technologies can be said to include a broad array ranging from 'old' technologies such as radio and TV to 'new' ones such as cellular mobile communications; while networks may be comprised of copper or fibre optic cable, wireless or cellular mobile links and satellite links. Equipment includes telephone handsets, computers and network elements such as base stations for wireless service; while software programmes are the lifeblood of all these components, the sets of instructions behind everything from operating systems to the Internet.

European Commission, 2001

Placed within the wider context of eDemocracy, electronic engagement can be represented as a subset of a wider range of activities occurring at the intersection of public policy and new communications technologies (Figure 2). A wider range of case examples which fit within this area of activity (including relevant strengths and weaknesses) is provided in detail in Appendix B: Catalogue of eEngagement Models.

In Figure 2 we see a distinct emphasis on community participation in established (or emerging) policy processes where specific outcomes (e.g. decisions or programmatic implementations) are emphasised.

Figure 2: eEngagement as a Subset of eDemocracy

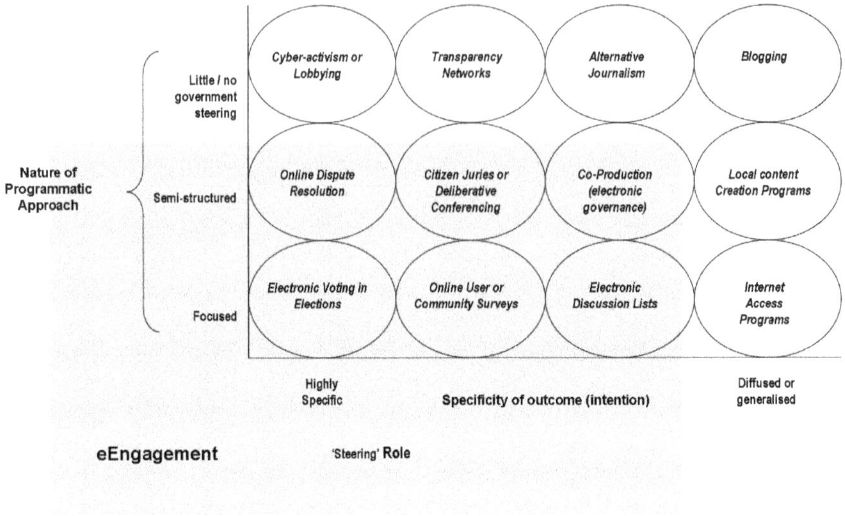

The advantages of this focus are:

- the instrumental nature of eEngagement is of direct value to policy managers – the investment of public resources is married with the objective of quantifiable outcomes in terms of improved policy development and greater community participation in decision-making;
- eEngagement activities often provide clearer means of program evaluation than the more diffused areas of eDemocracy activity, which either lie largely outside of government, or have multiple policy impacts that are often difficult to enumerate or measure (such as the democratic value of community content development policies);
- the approach focuses on issues of participation and public trust in government, allowing public sector managers a dedicated space for addressing the issues of democratic renewal through targeted activities that match their specific areas of policy responsibility; and
- it allows for a clearer delineation between different approaches to managing wider eDemocracy activities. In particular, capacity development and active listening approaches tend to have distinctly different management requirements than project-driven eEngagement activities.

2.3. Three Management Approaches

Based on these definitions, three different managerial approaches to implementation and management can be identified, each reflecting:

- different types of technologies involved;
- degree of complexity in program delivery;
- objectives (specific / diffused); and
- process timeframes and the transition from *project* to *passive* approaches to *eDemocracy* (see Section 2.3.4).

The approaches characterised in this guide are:

- the *active listening* role as a passive form of management;
- the *cultivating* role focusing on capacity-building and the stimulation of action by others; and
- the *steering* role, being a programmatic approach with high levels of management and control.

2.3.1. Active Listening

The desire by some governments to present themselves as technologically advanced and responsive to the community has tended to lead to situations where electronic democracy is interpreted as a 'thing' to be delivered to the waiting (passive and presumably grateful) public.

During the late 1990s this was reflected in a tendency for governments to formulate specific eDemocracy policy statements combined with a number of high profile activities. The best example of this approach can be seen in the United Kingdom under the early period of the Blair Labour government.

This can be beneficial in advancing the eDemocracy agenda. However, the approach can be seen to assume that ICTs are a 'push' (one-way) medium like television in which information is formulated centrally and then delivered to a passive audience.

The interactive nature of new digital technologies means that one of the important characteristics of the technology is the open participation by citizens and stakeholders in discussions of public interest. These discussions can include:

- unstructured conversation on email lists, through chat facilities, or on bulletin board systems (for example Yahoo! Groups; http://groups.yahoo.com/);
- expression of public opinion through alternative and non-profit online news publications (such as the OnLine Opinion magazine [http://www.onlineopinion.com.au/] or more specialist internet media); and
- the increasing number of 'citizen journalists' publishing on personal websites, blogs, or syndicated multimedia (podcasting or video blogging).

Listening management approaches are common throughout the public sector to allow for quick reactions to emerging issues or problems. This is particularly so amongst policy officers who are routinely tasked with monitoring mainstream media on behalf of their agency and Minister.

While this 'listening' is often undertaken in a relatively *ad hoc* manner, the inclusion of ICT-based listening approaches can be useful in that:

- information can often travel through electronic networks much faster than conventional media, thereby offering the potential for increased responsiveness;
- there is a range of commercial and free services [3] that automatically identify key terms and phrases from established media and alternative media and provide instant, or periodic, updates; and
- the introduction of RSS-type subscription services [4] allows for the customisation of news and information aggregation via desktop and mobile software.

While some might assume that a *listening management approach* is a euphemism for inactivity, an *effective* listening approach does require specific planning and management. Active listening requires:

- an investment in time to undertake *environmental scanning* to identify important sources of information. These sources need to be refreshed and renewed on a regular basis;
- a specific allocation of staff time to the collection of information (monitoring);
- establishing a mechanism by which information can be stored, searched, indexed, retrieved and interpreted in a meaningful way; and
- some means of establishing and assessing the value of the investment in active listening, either for the purposes of appropriately valuing and rewarding staff time, or as a mechanism for justifying this activity given its relative opportunity cost. One of the ongoing concerns associated with this form of eDemocracy activity can be the high 'noise to signal' ratio, being the poor return in terms of valuable information that can be gathered given the investment of time required to sift through irrelevant, uninformed, or misleading views and opinions.

---

[3] For example Google Alerts (http://www.google.com/alerts) for online news or Technorati (http://www.technorati.com/search/) for blogs.
[4] RSS (Really Simple Syndication) is a type of Internet file format that allows for information to be aggregated through the selection of a range of 'feeds' that are often updated by online publishers. These could include formal news services (the New Zealand Herald, for example, offers standard and customisable RSS feeds from its website, see: http://www.nzherald.co.nz/index.cfm?c_id=1500921&ObjectID=10125125) and most blog providers offer RSS capabilities as a standard part of their online publication.

Regardless of these concerns, listening approaches can be valuable precursors to the introduction of more structured eEngagement processes. They can provide the means for understanding the existing tenor of conversation, collecting useful background information and identifying elements of a policy issue that may be particularly engaging to the public.

It is entirely possible that key decision-makers in government will increasingly be as attuned to *blog* and website discussions of policy as they have traditionally been to television, radio and newspaper reporting.

Listening approaches are often employed *following* the conclusion of more structured eEngagement processes, either as a means of establishing popular views about the outcomes and impacts of policy decisions, or where the formal process has stimulated an active group of engaged stakeholders to oversight policy implementation.

Exhibit 9: 'Mass Listening' as Passive eEngagement Management

Elizabeth Richard of the Public Works and Government Services agency of the Canadian federal government notes that the internet provides public sector managers effective and interesting 'mass listening' tools. The proliferation of non-government, public email discussion lists on policy issues can give public sector managers interested in alternative views on policy and program implementation, avenues to undertake informal and unstructured listening to public views without necessarily engaging in formal consultative processes in the first instance.

The benefit of this approach lies in:

- the capacity to gather information informally, without the pressures of specific consultative timeframes;
- the ability to identify potential participants in formal consultative processes;
- hearing relatively candid points of view which may not be the same as arguments put in formal submissions – particularly where an issue is contested;
- the ability to absorb the level of debate (complexity, language used, degree of public understanding of policy issues) to allow public documents to be pitched at the right level;
- relative anonymity ('lurking'); and
- the ability to manage information gathering, particularly where there is concern that public consultation will lead to a large number of submissions (volume management).

2.3.2. Cultivating

Like the listening approach, cultivating or 'facilitative' management approaches rely on utilising existing skills found in civil society as the basis for successful community participation. Whereas active listening approaches can be valuable where there is an identifiable community of interest around the issue of concern, 'cultivating' recognises the need for outside assistance in stimulating participation.

In many policy areas, it may not be possible to identify existing communities of interest with which to engage. The public sector manager may find that the target audience lacks the technical capacity to use ICTs to participate in policy debate (where interested stakeholders are diffused through the society), or there has not been a recognition of a shared issue or concern that has given rise to mobilisation of interests.

Cultivation requires a number of activities:

- the identification of a specific and definable community of concern based on *locale* (such as a local community that has high levels of unemployment or crime) or non-geographic factors such as shared experience, or other identifiable characteristics (e.g. during 2005 the Victorian Office of Women's Policy undertook an online consultation associated with the experiences of working mothers across Victoria);
- definition of the characteristics of particular problems, which may be specific (lack of access to public transport, for example) or generalised (such as issues associated with school retention rates);
- determination of required inputs to address issue(s) of concern;
- development of participatory structures to deliver the required solutions;
- stimulation of collective activity; and
- development of the skills required to manage within the community (including appropriate governance and reporting requirements).

Depending upon the nature of the specific area of concern, the level of community involvement in initial planning and preparation may be limited or specific. This will depend on the nature of the problem and the existing capacity of local individuals or organisations to participate in early planning processes.

There are distinctly different approaches to 'cultivating' community participation, depending on whether:

- there is a clear recognition of a specific deficit which needs to be countered (the 'provision' model); or
- the community (geographical, policy, or community-of-interest) is active in defining the need, for example, customising a specific response to a social concern (the 'partnership' model).

The exact character of the response by the administering agency or agencies (cultivating models often necessitate partnerships across government) can be highly programmatic in character, or may be more intangible. Some programmatic examples include:

- the provision of ICTs (hardware);
- skills development;
- community training programs; and/or
- volunteering schemes.

It is also important to consider that less formalised activities can also fall under this approach. A good example is capacity-building in community groups that results from their inclusion in consultation and management processes. Inclusion enhances the position of organisations, thereby encouraging growth in membership and enhancing their representativeness. The result can be a stakeholder group of greater value to the public sector manager.

While these approaches can be used cynically,[5] they can be powerful in stimulating active organisations outside of government. Developing long term relations with formative groups can be important for the public sector manager with a medium term objective of creating a future partnership.

Given the nature of this type of management process, cultivation generally focuses on 'before and after' comparisons to determine measures of public value. For some projects this can be quite crude (e.g. percentage of free access terminals *per capita*) and others more complex and sophisticated (e.g. measures of social inclusiveness or similar 'social capital' metrics[6]).

Often, the key issues associated with *cultivation management* relate to the capacity to assess changes over time, particularly where programmatic activities have concluded, but there is an expectation of ongoing value creation.

2.3.3. Steering

In contrast to the above approaches, the final type of management response – steering – reflects a far more instrumental project management approach to policy delivery. *Steering* management approaches are common in developing eEngagement projects because of the emphasis placed on delivering short-term, specific and instrumental (policy development, acceptance testing and decision-making) outcomes.

---

[5] Such as 'licensing' a passive or supportive stakeholder group to the exclusion of more critical organisations.
[6] Defined by the OECD as 'networks, together with shared norms, values and understandings which facilitate cooperation within or among groups'.

Exhibit 10: Cultivating Approaches to eEngagement Management

Cultivating management approaches can yield powerful outcomes in the areas of community development, capacity building and the stimulation of active communities of interest.

Examples of this type of approach include:

- The Argyll and Bute Council of Scotland introduced a number of community telecentres in three remote island communities (Islay, Jura and Colonsay) offering personal computers with internet access and videoconferencing. The services have been highly popular, particularly during harsh winter months, with the services used to facilitate business operations, provide personal access to medical consultations (eService outcomes) and have been used extensively by the farming community to lobby the European Union over farm tenancy issues. While some of these applications were planned and expected by project managers, the use to which the videoconferencing service have been employed have been wider than expectations, leading to a multiplier effect of the technological investment.
- The New South Wales government established the *communitybuilders.nsw* website as a centralised clearing house for information associated with social, economic and environmental renewal through community-based organisations, non-profit groups and *volunteering* projects. The website provides information about organisation and management, financial assistance and planning and includes an extensive online discussion forum where people involved in these areas can exchange information and advice. While the Department of Community Services hosts and manages the website, the real value gained is through the interaction between citizens and citizens groups to solve local problems. See: http://www.communitybuilders.nsw.gov.au
- A variation of the communitybuilders model has been introduced by the British Broadcasting Corporation as its *Action Network* website (http://www.bbc.co.uk/dna/actionnetwork/). While *community builders* focuses on local renewal projects, *Action Network* has a more overtly political focus, allowing citizens to chat about political issues, start campaigns and network with like-minded individuals.

Definitions, Distinctions and Approaches to eEngagement

While steering approaches generally include participatory design elements appropriate to the anticipated stakeholder community, (either through the establishment of formal reference groups, or *ad hoc* consultation and negotiation), *steering* management approaches tend to be *agency*-driven.

This is due to the agency having:

- the capacity to develop a comprehensive engagement strategy;
- the resources to develop or acquire the appropriate technologies; and
- the ability to provide a 'hook' (access point) into the formal process of policy development in government.

Effective steering requires detailed preparation for the development of the eEngagement process, with clear process planning and well-defined timeframes. Flexibility in this approach is normally accommodated through reflective management and contingency planning. This is often important where the engagement process forms part of a specific policy initiative associated with the *executive*, or, where the consultation must meet the necessary timeframes for parliamentary reporting or legislative drafting.

The key aspects of appropriate steering management are:

- the integration of project development within wider strategic planning processes;
- the development of clearly articulated project deliverables, checkpoints and delivery timeframes;
- the need for specific program evaluation and reporting; and
- the tendency for these processes to be assessed against very specific outcome requirements (commonly expressed in terms of numerical metrics, such as numbers of participants, or output-based performance criteria).

Exhibit 11: The 'Electronic Discussion List' Model as eEngagement

The City of Darebin *eForum* pilot project in Melbourne reflects a conventional 'steering' approach to eEngagement management. The Council undertook to develop a structured online discussion forum which included Council staff and members of the community to discuss a range of local issues over a set period of time. Using basic email management technology, the council developed an engagement and promotional plan. A project officer recruited from local community groups moderated and summarised discussions and fed information collected back into the policy-making officers and Councillors at the end of each structured discussion. This approach was highly programmatic in character, with clear timeframes for action, close management of activities and control of interaction through the process of moderation.

2.3.4. Relationship Between the Three Approaches

While eEngagement activities tend to focus on *cultivating* and *steering*,[7] it is highly likely that a single project may require a number of different management approaches at different points of the planning and implementation process. A clear recognition of the relationship between project initiation, development, implementation, evaluation and closeout stages of any eEngagement activity can be extremely valuable in allowing the management group to recognise the appropriate management style for the particular phase of activity.

In addition, some reflection by project team members on their particular strengths and preferences can be useful in managing the transition between management approaches appropriate for different phases of project implementation.

Figure 3: Managerial Approaches Over an eEngagement Implementation Lifecycle

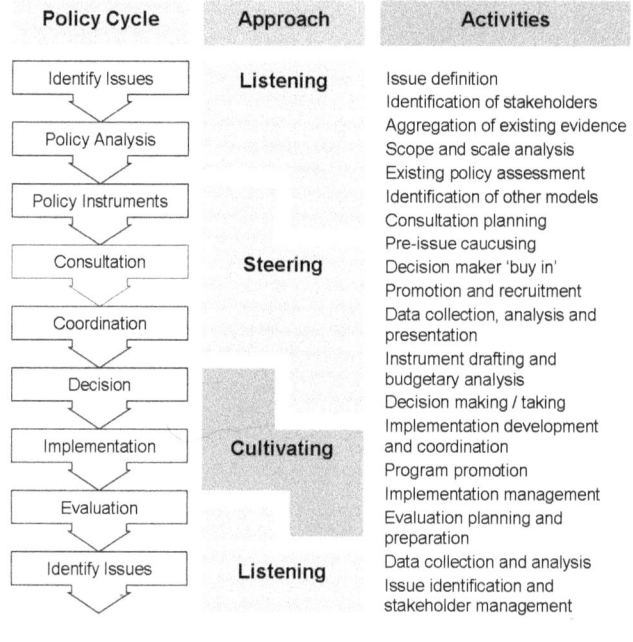

Adapted from: Bridgman, Peter and Davis, Glyn 2000, *The Australian Policy Handbook*, Allen & Unwin, St Leonards.

---

[7] For a more detailed discussion of different public service responses to the information age, see *Public Policy Forum 2003*, 'Archetypes of the Network Age: Articulating the New Public Service Reality', *The Public Policy Forum*, Ottawa, <http://www.pwgsc.gc.ca/archetypes/text/publications/report-e.pdf>

Managers who can recognise their preferred approach, or particular area of competency, are more effective at managing complex project implementations where a range of management styles are required. In some cases this may necessitate different members of the management team taking the lead role at different points in lifecycle of a project.

For example, Figure 3[8] presents a hypothetical eEngagement process that conceptualises the relationship between stages of the policy cycle and the range of different management approaches.

2.4. eEngagement and Electronic and Online Service Delivery

One of the ongoing debates within the literature on electronic democracy and engagement relates to the relationship between government electronic and online service delivery projects and political participation activities. Authors in this area consistently observe a *lag* between the work undertaken to place government services online and the use of ICTs in facilitating democratic participation.

Three hypotheses have been offered to explain this gap:

- *evolution:* that 'simple' transactions will be developed and implemented first (such as payment systems, bookings services and the like), with complex or 'messy' transactions and processes following;
- *anti-democratic:* that this reflects a lack of willingness on the part of government to be open and participative and is part of a broader malaise in liberal democracies. Authors in this area point to developments outside of government as better indications of the 'popularity' of the notion of electronic democracy, such as online protest movements and non-mainstream media; and
- *incompatibility:* that the processes are distinctly different and little can be gained comparing developments in one area with developments in the other.

All of these perspectives have some value and we can point to examples that illustrate each of them. However, it must be recognised that a simple delineation between the 'political' and the 'administrative' is an analytical fallacy that is undermined by observation of practice. A classic example is the provision of departmental and agency information online: a public sector activity that provides a useful public service (allowing greater access to government programs by members of the public) and also allows for greater transparency for democratic oversight.

---

[8] Adapted from: Bridgman, Peter and Davis, Glyn 2000, *The Australian Policy Handbook*, Allen & Unwin, St Leonards.

Exhibit 12: mGovernment

mGovernment' or 'mobile government' is the use of telecommunications technologies in the administrative process of government. With the growth of wireless telephone and internet access, increasing numbers of citizens are conducting their business, personal and government transactions using devices like mobile telephones, 'smartphones', wireless laptops and personal digital assistants (PDAs). These devices can be employed to access information services (such as telephone information lines or internet browsing) or conduct transactions online (book and pay for services, make appointments, complete forms and other regulatory requirements) and reflect the growing flexibility of people's employment and work / life balance. The next generation of mobile telephones (3G), for example, feature high-speed internet access that allows for the transmission of video (send and receive).

Like eGovernment, mGovernment has both internal and external applications. Inside the Public Service, techniques like teleworking allow greater employment flexibility, or the provision of portable computers allows for:

- 'smart' fieldwork which optimises time spent in the community and reduces the need for a return to base (such as in the areas of Policing, Community Services and some regulation and licensing areas); and/or
- home-based employment arrangements that allow for greater employment flexibility, staff decentralisation and reduction in the need for work-related commuting.

Externally, governments are looking at ways that these devices can be used to transact business with government (such as in remote service delivery) or means to 'push' information to members of the public (such as the use of SMS notification services).

The benefits of mGovernment are:

- increased flexibility of employment;
- greater reach of government information and service functions into the community;
- increased convenience of access to government; and
- increased choice of interaction.

The risks of investing in mGovernment lie in:

- further loss of interpersonal contact within the public service and between government and the public;
- telework 'bleeding into' personal life;

- reduced professional contact for teleworkers; and/or
- unclear development path for mobile technologies (questionable levels of uptake of advanced devices).

2.4.1. eGovernment Catalysts for eEngagement

While it is clear that the introduction of electronic and online service delivery infrastructure within the public sector provides a useful platform for developing eEngagement activities, it is useful to reflect on the relationship between these two areas of activity across four dimensions.

First, service recipients' experiences with electronic and online service delivery applications using ICTs closely resemble eEngagement projects associated with highly focused data collection. There is negligible difference between this and normal market research undertaken by government. The defining characteristic is the selection of participants based on their use of a specific service channel.

Second, where the objectives of an eEngagement activity are diffuse and the process of engagement is either semi-structured, or un-structured, in nature, it is possible to recognise a significant difference between these types of online transactional systems and conventional electronic commerce technologies, which tend to be based on highly specific and relatively rigid transactional process models, with limited capacity for members of the community to vary from the imposed structure.

Third, the electronic service programs of government can provide opportunities to expand eEngagement. This can be achieved through ensuring that the development of new service channels have the capacity to include consultation and participation activities. A good example of the possibilities here can be seen in the use of service delivery terminals for public consultation, particularly where the consultation focuses on issues of place.

Fourth, there can be opportunities for policy managers to provide significant input into the development of service delivery technologies to provide more policy-oriented user information from these systems. Electronic and online service delivery systems are commonly developed with the intention of introducing efficiencies or extending the reach of public services and these projects can focus only on highly 'rational' outcomes (for example, new systems are developed only to introduce cost efficiencies in existing business practices). Given the often considerable investment of public money in the development of these technologies, consideration of system development that allows for the capture of information for policy analysis can provide significant benefits to policy outcomes. These benefits can include:

- the identification of specific user groups (and, by extension, under-represented groups);
- uptake rates for new programmatic offerings (such as time taken browsing basic information regarding service offerings versus time spent undertaking transactions);
- recruitment of participants for ongoing consultation processes or subscription to news and information services; and/or
- polling on issues related to the specific transaction, or of relevance to the type of user (e.g. associated with a different policy issue).

In addition, it must be recognised that one of the most powerful aspects of electronic and online service delivery is the capacity for information to be captured, analysed and presented in *real-time*. This aspect of eGovernment can represent one of the most powerful opportunities for public management.

2.4.2. Difficulties and Tensions

Public sector managers with an interest in eEngagement can play important roles in the development of electronic and online service delivery activities. However, it is also important to take into consideration the *business culture* of the business units tasked with developing the systems. Indeed, business units will require considerable persuasion to incorporate 'fuzzy' or 'soft' processes and capabilities within their business systems.

Where the eEngagement team is attempting to *piggyback* on a hardware installation, (e.g. accessing participants via a service kiosk, where access may be rationed due to scarcity), the justification required to argue for the integration of an eEngagement initiative may be considerable. These difficulties can be particularly acute where:

- the business units are culturally or structurally removed from policy staff and engagement activities or priorities; and/or
- where the transactional service is built on a highly secure platform (such as one based around a payment gateway).

> **Exhibit 13: Relationship Between mGovernment and eEngagement**
>
> mGovernment is compatible with eEngagement, but has implications for public sector managers investing in these concepts:
>
> - Public servants need to consider the range of devices used by members of the public to interact with government information services. The appropriate design of websites, for example, can allow for ease of access by members of the public with devices that have small screens and low-speed internet access. Alternatively, information stored online can be re-purposed for use in Interactive Telephone Services;
> - Information-on-demand permits timely participation in government consultation processes. The Queensland Government's *Generate* youth service allows for SMS messages to be sent to subscribers notifying them of new consultations; and
> - Portable ICTs permit a range of possibilities, from simple participation to remote data collection. For example, the increasing prevalence of Global Positioning System (GPS) location data has been used in the United States to encourage the creation of local pollution maps by volunteers.

2.5. The Digital Divide: An Absolute Barrier?

A common concern regarding the adoption of eEngagement initiatives is the limited use of ICTs in the wider community. With approximately three quarters of the New Zealand and Australian populations using the internet relatively frequently[9] the level of use of this technology is far from the near universality of other communications appliances like telephones.

The gap between universal access and the current penetration of ICTs is commonly referred to as the *digital divide* and represents a real concern for policy makers as it represents a different form of non-participation, namely, non-participation in the information society/economy.

It can be argued that this divide limits the value of new channels for engaging the public in policy processes. As specific segments of the community are excluded from these technologies, the results of using eEngagement are systematically skewed, particularly excluding people who are considered to be generally under-represented in conventional policy processes, such as the poor,

---

[9] Up to date statistics on Internet use, particularly by location and frequency, are not presently available and these figures are based on estimates only. During 2006, both New Zealand and Australia held their national censuses, the data from which should be systematically released by both national statistical agencies from early 2007.

migrants, indigenous people and those with limited educational backgrounds. eEngagement can be seen as anti-democratic leading to increased access by people in the community who are currently 'well served' by existing democratic structures.

2.5.1. Nature of the Divide

While this concern has relevance and is worthy of serious consideration at the initial stages of eEngagement project development, it does tend to promote a simplistic view of the average user of new communications technologies as:

- white
- male
- urban
- 25 to 40 years of age
- professional
- university educated

While this might have been an accurate portrait during the 1990s, the uptake of ICTs throughout the community has developed in unexpected ways. These include:

- the rise of 'silver surfers' – retirees who find email and the internet an interesting and rewarding past time and means to maintain contact with children and grandchildren;
- the use of ICTs in some migrant communities to access international news in their preferred language and maintain familial and business contacts in their country of origin;
- the use of the internet in rural communities, either through the emerging area of 'teleworking' (remotely working from home) or farm-based ICT use to engage with world markets and use advanced sensing technologies (such as digital dam level indicators and remote cameras);
- different usage patterns for similar technologies between age groups (e.g. youth versus business mobile telephone use); and
- the significant narrowing of the gender gap.

> **Exhibit 14: Mobile Phones Buck the Digital Divide** [10]
>
> While the rate of internet adoption has slowed over the last five years, the penetration of mobile telephones in Australia and New Zealand continues to be strong. Both nations approach near 100 percent penetration of this technology and users are increasingly comfortable engaging with interactive services using mobile telephones.
>
> In 2004-05 it was determined that 38 percent of Australians over the age of 16 had used their phone to participate in a competition via SMS.
>
> Telephones exhibit a faster adoption curve (both market penetration and uptake of new features) because:
>
> - they have a short lifecycle (they are replaced more frequently than computers);
> - their total cost of ownership is low and their cost can be deferred over their operating life (the handset cost is often integrated into service costs); and
> - they are comparatively simple to use.

In addition, a large number of government and not-for-profit programs exist to improve access to ICTs by under-represented target communities, either through subsidised purchasing schemes, or through the provision of public access terminals in community centres, public housing estates, schools and job service organisations.

Despite these initiatives, the problem of the digital divide persists. During the initial popularisation of the internet in the mid-1990s, when growth rates for ICT usage were very high, the digital divide was characterised simply as an effect of the combination of technological diffusion speed and cost barriers to adoption. The assumption was, at this time, that as the number of users embracing the technology increased, more commercial vendors would be encouraged to enter the market, resulting in an easing of cost barriers. Although increased demand has driven costs down, this has not been enough to close the digital divide. In fact, adoption rates have slowed and some communities have shown limited uptake of ICTs.

The reasons for the digital divide are complex and not easily addressed by policy makers. They include:

---

[10] Niesche, Christopher 2006, 'Government Must Free Mobile Market', *New Zealand Herald*, 24 July; Fisher, Vivienne 2005, 'Australians Embrace Mobile Phones', *Australian PC Authority*, 31 May; Nielsen//NetRatings 2005, *The Nielsen//NetRatings Australian Internet and Technology Report 2004-2005*, Nielsen//NetRaitings, New York.

- a higher price 'floor' arising from the need to acquire and maintain both ICT equipment (with rapid replacement requirements due to obsolescence) as well as access accounts (often *in addition* to existing communications costs);
- lower levels of competition for some data services than anticipated, due to limited competition in the provision of network infrastructure (particularly outside of urban areas and in the wholesale market);
- difficulties in moving some parts of the community online (particularly those without full-time employment, with poor English language skills and older citizens);
- 'transitional' delays, as users move between older and newer technologies, or basic versus advanced services (e.g. dial-up to broadband, 2G-3G mobile telephony); and
- higher than expected barriers to entry. This is due to a combination of low technical literacy levels in parts of the community and the rapidly changing technical environment (making the 'cost' of maintaining accurate technical literacy high – this has been particularly exacerbated by socially-undesirable activities online that are not well regulated by national governments[11]).

2.5.2. Implications of the Divide

The use of eEngagement systems will include (or be included within) a broader strategy that includes conventional 'offline' means of participation. For simple engagement approaches (such as the solicitation of submissions or surveying), this may simply require the provision of paper versions of discussion documentation and postal response mechanisms, whereas, for more complex processes (particularly deliberative ones or where specific sampling rules are applied) this may mean running parallel processes.

Where parallel processes are conducted, the managerial implications may be significant. These can include:

- issues of timing: often on- and offline processes work on different timescales and synchronising parallel processes can be difficult to manage;
- issues of comparability: for parallel processes to work they need to be similar in scope and interactivity. Where complex ICT applications are employed, determining how the richness of online eEngagement can be mirrored offline may be difficult; and/or
- separate or integrated discussions: if there is a desire for 'cross talk' between the on- and offline communities, then consideration is required about how this will be managed. This may be a significant issue where there is a conscious desire for information, or experience, sharing between these two

---

[11] Such as the proliferation of *malware* (virus, *spyware*, or *Trojan-horse* software) requiring user vigilance and the growth of *SPAM* and online fraud (*phishing*, identity theft).

groups (especially where the composition of the on- and offline groups is distinctly different).

2.5.3. Beyond the 'One Divide'

While these issues can be seen as daunting, it is important to conceptualise the digital divide as one of many different and overlapping, barriers to participation. While ICTs can provide enhanced access to policy processes for *some* (and can, therefore, be seen as democratically problematic), they also can be used to overcome other access problems.

Figure 4 illustrates a range of divides that overlap and provides insight into how a mix of ICT-based engagement and conventional approaches can create better overall outcomes in the reduction of barriers to participation.

Figure 4: Digital Divide or Multiple Divide?

Divide	Description	ICT Implications
Bandwidth	Access to ICTs, but slow access speeds. May be because of poor infrastructure, old equipment, remoteness, basic ISP account, or the use of technologies like 2.5G mobile telephony	Necessity of design of eEngagement for low-bandwidth environments
Digital	Lack of access to ICTs, either because of cost, skills, interest, language, or infrastructure	Importance of offline complementary processes, or provision of ICTs as part of eEngagement strategy
Educational	Limited education can limit access to policy processes through limited capacity to engage with briefing materials, low understanding of government processes / structures	Use of ICTs to education (primers, simplified language, etc.)
Linguistic	Poor / no English which limits access to formal consultation documents	Provision of translations or spoken equivalents
Mobility	Limited capacity to travel to physical venues, either due to poor transport infrastructure, limited financial resources, career status, or physical impairment	ICTs to overcome distance issues
Motivation	Lack of interest in issue, limited belief in value of participation, disenchantment with process	Use of engaging content, demonstration of commitment through activity
Time Poor	Limited ability to commit blocks of time to ongoing processes. May be due to career commitments, working hours (long, non-standard, erratic or on-call), or parenting	Use of asynchronous communications to manage time constraints
Vision Impairment	Vision impairment can be a barrier to participation where process is conducted via printed mail, are advertised in conventional printed matter only (newspapers) or where participatory forums make heavy use of visual aids (PowerPoint-type presentations)	Provision of material in digital form, use of spoken word versions, distribution of printed matter equivalents prior to physical meetings

Through a broader conceptualisation of the community's access difficulties, we can achieve a better understanding of the appropriate role for ICTs in engagement processes. In addition, where ICT access barriers can be seen as disproportionately associated with some groups in the community, we need to be cautious about *universalising* this assumption. Where the approach taken to the eEngagement process is based on sampling to develop a representative section of the wider community (a cluster or quota sampling methodology) lower levels of ICT uptake in some areas of the community can be recognised and addressed through the use of appropriate quotas and additional recruitment in areas of under-representation.

Recognising areas of low uptake through eEngagement program design and implementation can be a catalyst for partnering with other community access programs. One of the key lessons learned during the last decade is that digital divide issues are often most effectively addressed through a combination of technical access provision, training and the incorporation of relevant compelling content. eEngagement activities can be seen as a highly effective way of motivating participation in the information economy.

3. Designing the Right Approach

While the great advantage of eEngagement is the wide array of approaches that can be applied to resolving policy and participation issues, the disadvantage of this flexibility lies in the difficulty of determining an appropriate approach at the outset of project initiation.

Technologies like the *World Wide Web* draw their power from the vast array of applications to which the basic technology can be applied. However, the very strength of this technology can lead to 'option paralysis'. In such cases, determining an appropriate model or program design from the existing case examples, or choosing from the vast array of potential and hypothetical applications that are still being developed, can be extremely complex.

One of the problems associated with eEngagement activities to date has been the tendency to emulate a limited number of options, rather than engage in broader experimentation and refinement. In particular, electronic discussion forums have been among the more popular eEngagement approaches.

It is possible here to recognise a limitation associated with 'unreflective' policy transfer and 'lesson-drawing' across jurisdictions. The important thing to consider when examining successful case examples from different cultures, jurisdictions, or policy areas is that:

- the nature of public participation in particular policy areas is highly variable, even within a local area. Careful consideration of the history of participation in your area of concern will guide your development process over the views of 'experts'. The relative newness of eEngagement does mean that expertise in this area is in short supply; and
- the political culture of a jurisdiction is often highly influential in determining the nature and extent of public participation in policy processes. In addition, the way in which public participation activities are assessed is highly dependent on local notions of what 'democracy' is. In areas where participation is high, the development of an eEngagement activity is likely to focus largely on the design and implementation of the communications channel, whereas in areas where participation is low, has not been encouraged in the past, or where high levels of community cynicism exist, a large amount of the work in developing a new engagement strategy will be focused on issues of community development and the creation of trust.

The basic design of an eEngagement approach will be highly influential in determining the success or failure of the final process. Maintaining commitment and support for eEngagement initiatives over time – particularly among key decision-makers – requires careful consideration of the following essential matters:

- what shape has the engagement process taken and is it appropriate?
- are public and/or target community expectations about participation realistic?
- are projected outcomes realistic?

3.1. Key Decisions

During the initiation phase of project proposal planning it is important to consider the following six questions:

1. What is the issue(s)?
2. Who are the audience(s)?
3. Consultation versus collaboration?
4. What objectives do we have for this activity?
5. How interactive will this process be?
6. Which is the right channel (communications technology) to use?

In a normative sense, the ability to define and answer these six questions will define the minimum requirements for project approval within the authorising environment. In addition, the appropriate articulation of answers to these questions (and consideration of any complexity, or contradictions, presented by the answers) will provide a solid foundation for the development of an effective implementation plan.

3.1.1. What is the Issue(s)?

The introduction of any engagement practice (on- or offline) will be predicated on the clear articulation of any information deficit(s) within the organisation, or government, as a whole. However, this is not always the case, even though it is essential to any consideration of the shape and nature of the issue, or issues, under consideration and will be instructive in shaping an appropriate eEngagement strategy.

Three examples of areas requiring careful consideration[1] of the policy issue under review include:

- where the introduction of eEngagement is being specifically undertaken because of limited public participation in the work of an organisation generally. In this example, it will be important for the project team to ascertain the range of policy questions that will form the subject matter for ongoing consultative processes. A good example of this is the introduction of centralised institutional online consultation tools (where the objective is to establish an ongoing online community or reference group for the organisation), with the expectation that a series of policy issues will be fed through this mechanism on an ongoing or regular basis; or

---

[1] Through techniques such as concept mapping, brainstorming, or idea writing.

- where the particular policy issue under consideration has been handed to the agency by a superordinate body and the agency has been tasked with the process of implementing a consultation process, but has limited control over the capacity to define the issue under review. In this example, the implementing agency may conclude that the issue has greater levels of complexity than the scope of the brief. In such cases, if the agency is unable to receive authorisation for an adjustment to the issue definition provided, (which may be the result of a Parliamentary reference, stem from a electoral commitment or reflect party policy, or due to the proclivities of an individual manager or Minister), the inherent limitations of the issue definition will have to be countered by careful consultative design. A classic example of this problem is where superficial phenomena are specified for consultation, but the agency has been forbidden to examine deeper causal, or contributing, factors due to political sensitivities; or
- where the nature of the issue itself will – to greater or lesser degree – determine the form taken by the consultation process. Examples include:
  - where the particular area of public policy has been captured by a small 'insider' group and there is explicit recognition that participation needs to be broadened through a mechanism which limits the capacity of one organisation to dominate the participatory process;
  - where the policy area has been subject to considerable disputation leading to the development of a polarised set of stakeholders and there is a desire to reconcile these tensions;
  - where the Executive has specified a particular outcome for the process, such as clearly articulated organisational mission statements, community-designed performance targets, or a single jointly-authored final report; or
  - where there are low levels of public understanding of the issue, its nature and/or causal factors and the participatory process necessitates a period of community education prior to data collection or deliberation.

3.1.2. Who is the Audience(s)?

The nature of a policy issue will define the audience or audiences for any engagement activity. A clear understanding of the policy issue can assist in the indentification of a target group and assist understanding of the characteristics of the target audience, leading to:

- a clear understanding of the likely expectations of the target audience in the participatory process;
- their level of understanding of the policy issue;
- appreciation of the issue background, including areas of tension and conflict which need to be presented in context and possibly with care and attention to partisan sensibilities;

- an appropriate approach to *message development,* possibly requiring direct contribution from competing stakeholders;
- some indication of preferred communications channels to reach the audience(s); and
- likely responses to different models of consultation.

In many circumstances, this will simply require an appropriate period of reflection about the characteristics of groups with whom the agency has had interactions in the past, while some consultative activities may require that specific market research be undertaken to develop an appropriate classification ('market segmentation' approach).

Key questions to consider are:

- is there a single community that can be defined as having internal coherence (e.g. members of the group consider themselves to represent a community)?
  - if so, what are the characteristics of this community?
  - if not, how will the various types of stakeholder groups be classified?
- are there particular members of the community or stakeholders who could be identified as 'opinion leaders' or 'influentials'? [2]
  - if so, what role will these individuals play in the consultative process (do they expect to play a positive or negative role, can they speak for wider sections of their community, will they be included or excluded at all)?
- is there a single unit of analysis with regard to stakeholders to be consulted (individuals, family groups, interest groups, businesses, etc.)
- is there a latent group, [3] or groups, who may be affected by the policy issue, or potential policy responses, which will be incorporated in the consultative process?
  - if so, what are the characteristics of this group and how can this audience be recruited into the eEngagement process given their latent status?
  - at what point will this group or groups be brought into the process (for example, if the policy issue only becomes relevant to them in the event that a specific policy recommendation emerges)?

---

[2] For an interesting discussion of the role of *influentials* online see: Institute for Politics, Democracy & the Internet, 2004, *Political Influentials Online in the 2004 Presidential Campaign*, Graduate School of Political Management, George Washington University, Washington DC, <http://www.ipdi.org/UploadedFiles/influentials_in_2004.pdf>

[3] Sometimes referred to as a 'potential pressure group'; Truman, David 1951, *The Governmental Process: Political Interests and Public Opinion*, Knopf, New York.

> **Exhibit 15: Recognising Different Audience Types in an Extensive Online Engagement Activity**
>
> In undertaking a wide-scale online consultation over the development of the new European Union Constitution, the project management team was careful to identify two types of audience:
>
> - individual members of the public participating in referenda, at the national level, to adopt or reject the final draft; and
> - interest groups and other organisations that have a stake in the development of the European Community. These included business organisations, non-profit groups, parties and *ad hoc* associations.
>
> The resulting system developed to undertake the online consultation process, *Europa*, included two parallel consultation processes, one for individuals and the other for organisations.

3.1.3. Consultation versus Collaboration

One aspect of citizen engagement activities that is of particular interest is the capacity for participation activities to go beyond pure consultation (data gathering and collection) to the empowerment of stakeholders and service recipients, thereby giving effect to greater, or lesser, control over the decision-making process. This is often characterised as a *conceptual continuum* between pure data gathering activities (consultative) and the complete devolution of authority to local communities (direct decision-making).

The desire for greater public control over the process is often predicated on a belief that this can result in either: (a) better decision outcomes; or (b) greater acceptance of outcomes achieved ('ownership'). This belief, which will be tested as part of the project preparation phase, reflects a recognition of alternative sources of expertise (local knowledge, 'folk' wisdom and experiential learning) over *classical* forms of expertise prized by policy analysts (theory-based knowledge, often entrenched within a scientific mode of inquiry).

This view is often articulated in conceptual models, such as the example provided in Figure 5, the articulation of a linear model starting with complete government control over the process of consultation and data gathering and builds towards complete devolution. As one moves down the spectrum, greater levels of community participation and decision-making are introduced.

Figure 5: The 'Consultation Continuum'[a]

↓ Increasing Participation	Consultation	Exclusive government decision-making, little consultation	Control Mode
		Listening, dialogue, limited consultation, no impact on decisions	Briefing Mode
		More open debate, shared analysis of problems, scope to influence decision-making	Debate Mode
		Joint agreement on solutions, strong potential to influence decision-making	Consensus Mode
		Joint decision-making with regards to implementation and policy	Coordination Mode
	Partnerships	Participation in design and delivery of programs and services	Operational Partnership
		Shared decision-making in policy development and program / service design and delivery	Collaborative Partnership
	Devolution	Transfer of responsibility for program / service design and delivery	Devolution

[a] Sourced from Clarkson, Beverley and Rigon, Joanne 1998, *Consultation Practices: Departmental Overview*, Department of Indian Affairs and Northern Development, Ottawa.

3.1.3.1. Implications of the Continuum

While this (and similar) 'consultation continuum' models[4] (see Appendix A: Policy Cycle Engagement Model) are valuable in allowing public sector managers to consider engagement processes that include greater control by participants over outcomes, this conceptualisation can be criticised for implying a normative progression from consultative processes, through to collaborative approaches, to direct decision-making by members of the consultative group.

In addition, while these models are useful in that they introduce the notion of direct and deliberative democracy to public consultative processes, (long overlooked and excluded from agency planning), it needs to be recognised that the movement from consultation to devolution, or local decision-making, involves both:

- a radical shift from traditional approaches to public engagement. The nature of the issue and audience, combined with the expectations of senior decision-makers, will be significant in determining the degree to which the engagement process is consultative or collaborative in character; and
- very different approaches to the management of the activity. As per the discussion in section 2.3, Three Management Approaches, the role of the delivery agency, as one moves from consultation to participation, progresses from a steering, to a cultivating and listening approach.

Often, the implication is that, as one progresses towards more participatory approaches, the level of 'democracy' increases. However, this can be problematic if the engagement process is not representative, or where deliberative processes

---

[4] For example see the simplified model in Coleman and Gøtze's *Bowling Together*. The full reference is included in *Further Reading*.

have the capacity to move significant sectors of the wider community to a new policy position that differs signifcantly from 'popular opinion'.

3.1.3.2. Reconceptualising Consultation and Collaboration

The 'consultation continuum' is also limited by its single axis of analysis. The degree to which participants can be active in making decisions regarding the consultative process needs to be considered when making choices about consultation versus collaboration.

Figure 6,[5] sets out a two axis model. The first axis takes account of the continuum illustrated in Figure 5, while the second introduces the notion of control over the interactive environment (the process by which the consultative or deliberative activity will occur).

Figure 6: Consultation and Collaboration

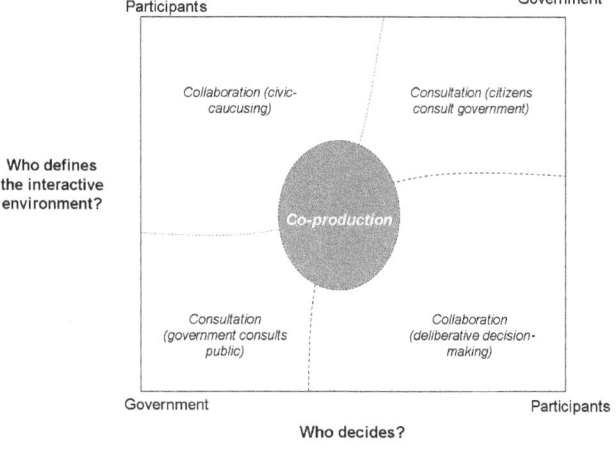

Adapted from Arnstein, Sherry 1969, 'A Ladder of Citizen Participation', *Journal of the American Planning Association*, vol. 35, no. 4, pp. 216-224.

Using this approach it is possible to envisage an engagement process that focuses on a range of consultation and collaboration aspects. This allows control, even when a key decision-maker or agency may be cautious about including significant *decision-making* capacity within the consultation process. In this case, the desire for community control can be incorporated through providing flexibility to participants to determine the *structure and processes* of the consultative environment. The advantages and limitations of this approach are illustrated in Figure 7. What must be remembered is that the selection of collaborative over

---

[5] Adapted from Arnstein, Sherry 1969, 'A Ladder of Citizen Participation', *Journal of the American Planning Association*, vol. 35, no. 4, pp. 216-224.

consultative approaches marks a departure from common practice and will normally require significant justification or motivation (commonly either a response to ongoing policy failure on the part of government, or the need for a 'clean slate' over contested issues).

Figure 7: Comparative Benefits of Consultation versus Collaboration

		Advantages	Limitations
Consultation	Government-led	Control over the process allows for an orderly collection of public points of view in a timeframe which can be identified and planned in advance. This is useful where the issue is both well-defined and requires a quick response from government	Unsuited to issues where the problem is ill-defined or where it is difficult to determine in advance how members of the community will respond to the request for participation. This approach is unsuited to consultative processes that require mediated outcomes, or have the objective of developing consensus or community-building outputs
Consultation	Citizen-led	Allowing participants to define the process by which information will be gathered for decision-making by government is a useful means to empower participants without the Executive feeling decision-making or control has been lost. This approach can be useful in developing trust and also motivating high levels of community participation in the process	Allowing community control over the development of the consultative process can create problems in three areas: (a) loss of control over consultation timeframe; (b) community may have limited expertise in organising participatory approach; and (c) can attract criticism of 'talking about talking'
Collaboration	Government-led	Allows public decision-making over aspects of policy preference within a structure controlled by the agency. This can be useful in ensuring that the process maintains timeframe and boundary controls (e.g. does not spill outside the agency's remit or area of expertise) while high levels of citizen participation are fostered	Agency can be criticised for manipulating outcomes by controlling decision-making process
Collaboration	Citizen-led	Can be highly democratic, the ability of the participants to define the process and outcomes can be very attractive and useful in redefining areas of particular contestation	Expensive, requires representative selection of participants and delegated authority for direct decision-making by key decision-makers in government (such as ministerial endorsement). Limited control over processes and outcomes can create the perception of significant risks, particularly if process becomes 'hijacked'
Co-decision	Shared control	Shared decision-making over process and outcomes by government and non-government participants can create balanced decision-making which incorporates the views of key stakeholders (desires) with an understanding of implementation capabilities, budgetary constraints, etc of public servants (grounded decision-making against capabilities). Can be useful for ongoing managerial relationships	Often difficult to establish (particularly to ensure a genuine balance of authority and decision-making between government and community), appropriate selection of community participants can be manipulated or contested, skills / interest in the relevant stakeholders community for ongoing participation may be limited or absent

In making the final selection of a preferred approach, it is important to consider:

- the benefits of delegation versus consultation, as per Figure 7;
- the willingness of Executive, or elected, decision-makers to delegate and maintain commitment to delegated outcomes, particularly on politically-contentious or difficult outcomes. This aspect of the choice of deliberative and collaborative approaches is particularly critical, as the failure to act on decisions made in collaborative processes can be particularly toxic to public trust in government;
- the effectiveness of the selected approach in shifting the level of the policy debate (tone, types of participants), versus other means to achieve these outcomes (widening public debate, using third party organisations etc.); and
- the response of existing 'insiders' or beneficiaries. One of the particular difficulties of engaging the public through more direct decision-making models is the reaction of insider groups whose responses can include boycotting and active resistance to the process.

3.1.4. Setting Objectives

Establishing documented (and published) objectives for the eEngagement process is a necessary primary step in the development of the implementation plan. Where possible, the objectives articulated early in the process will be the basis for evaluation measures used in the latter stages.

Objectives are likely to develop or shift over time, (particularly where the process is citizen-led, or collaborative across agencies). In general terms, however, objectives may be:

- broad in nature, possibly reflecting a 'community building' or 'visioning' approach, in which the development of a conversation, or dialogue, amongst participants represents an end in itself. Often, these types of objectives are the most rewarding form of eEngagement activity and result in a tangible and durable, outcome. That said, these types of projects also commonly require considerable planning and coordination, and can, therefore, be complex to develop and administer effectively. eEngagement activities that have broad objectives are at risk of difficulties associated with:
    - matching outcome metrics with performance;
    - maintaining momentum over time; and
    - ensuring ongoing support from senior management and Ministers, who are focused on electoral cycle timeframes; or,
- Narrowly-focused or *instrumental* in nature ('problem resolution') and, therefore, easier to develop and assess. Given the short-term nature of these activities, managers must be cautious of:
    - accusations of limited cost effectiveness;

- problems maintaining ongoing commitment to the eEngagement approach following project completion;
- difficulties maintaining staff and skills in the agency; and
- 'flash' effects (outcome measures) that have limited medium to long term public value.

The articulation of objectives is best undertaken in a clustered manner, particularly where objectives may not be equally amenable to later evaluation. While there is a common desire across government to focus on outcome (over output) measures of productivity and success, this is not always possible and well designed objectives will allow substitution of relevant activity measures for missing outcome data.

3.1.5. Degree of Interactivity

One of the most compelling characteristics of new media for the purposes of public engagement is the use of interactive technologies. Interactivity, in this context, refers to the ability of the consultation process (human aspects) or system (technological aspects) to respond to the actions of the user and vary from a heavily pre-determined process to one that is more flexible.

In this way, the notion of interactivity also has implications for any decision about whether to make a process consultative or collaborative. Interactivity can be defined as 'to act mutually; to perform reciprocal acts' and can reflect:

- the humanising aspect of technology: the way technology responds to the user in a dynamic manner, accommodating their input and/or preferences; and
- the nature of the eEngagement processes as flexible and user directed, or a pre-determined process.

The desire to create user-friendly technologies, together with the type of eEngagement processes being undertaken, will determine the level of interactivity required and desired. This, in turn, will shape the choice of technology used.

Most digital technologies are interactive to some degree. Even relatively limited channels (2G mobile telephone networks, interactive television, menu-driven 'static' DVD videodisks, etc.) provide a great deal of flexibility in terms of the interactivity inherent in the engagement process.

In the eEngagement process, it is common to think of interactivity as a technical characteristic. However, the interactivity of an eEngagement system is commonly the result of deliberative design to incorporate interaction between participants, either vertically (between the participants and the agency), horizontally (between participants), or both.

> **Exhibit 16: Comparative Interactivity of Two Online Petition Systems**
>
> A good comparative example of horizontal and vertical interaction is to compare the parliamentary petitions systems of Queensland (Australia) and the Parliament of Scotland. Both systems provide access to the traditional petition system of their Parliaments, where members of the community can express their view on any issue to the Parliament.
>
> However the two systems differ considerably in their level of interactivity. The Queensland system (http://www.parliament.qld.gov.au/view/EPetitions%5FQLD/) permits members of the public to sign the petition only – a singular direct interaction with the petition system. The Scottish system (http://epetitions.scottish.parliament.uk/) provides a similar mechanism, but also includes the ability for petitioners to discuss the issue amongst themselves through a moderated online forum that is part of the petitions website.

In thinking about what role interactivity can play in the development of the eEngagement process, consider five types of interactivity:

- *synchronous versus asynchronous* interaction: is the process to be conducted in real-time (live) like a public meeting (with the advantages of spontaneity) or will participation be staged over time (allowing greater participation for people without the ability to commit a 'block' of time, or allowing a more wide-ranging debate, discussion, or voting process?
- the number of iterations of interaction: what will the 'intensity' of the interaction be? Will it be a series of small interactions, or only a small number of more intense interactions over time? The advantage of the former approach is the extent, range and free-flowing nature of the process, the latter can exhibit a tendency for high-volume participants to 'drown out' other voices;
- the level of interaction between users: is there an explicit desire among participants to have a dialogue, or interaction, with each other? This may be useful where the desire is to stimulate (not lead) a discussion or debate, or in deliberative processes where the participants are tasked with presenting their points of view and arguing these towards a voting process. On the other hand, where there may be particular risks to participants (see section 4.6, Managing Risk) or, where data collection only is required, interaction of this type may be counter-productive;
- interaction between different channels: where multiple channels are involved, (or participants split between sub-groups, as in a deliberative conference), one of the core questions may be the cost-benefit of allowing discussion or interaction across these channels. In some cases this may be highly desirable,

(e.g. where there is a recognition that different channels will attract different types of participants and the motivation for the eEngagement project is inter-group learning). On the other hand, splitting groups into selected clusters, or teams, can be valuable in moving people out of entrenched positions supported by their peer groups and lead to deeper interactions between individuals;
- inclusion of elected representatives in the process: this is often one of the more difficult questions to resolve and is highly dependent on the nature of the issue under review and the hosting organisation. In many cases this may represent only token endorsement of the process (e.g. the Ministerial greeting), in others, elected representatives may be a core part of the process, using the project to supplement and enhance their interactions with the community. [6] On the other hand, some consultation units recognise risks in this option, including:
  - the prominence of the individual will detract from public participation;
  - the elected representative may dominate, or hijack, the process to promote a personal viewpoint, or reiterate party/government policy to the exclusion of debate; or
  - the process being used by a partisan group to 'score points' (e.g. through 'stacking' the forum).

Overall, three questions need to be answered:

- how much interactivity is *needed* ?
- how much interaction does the audience *want* ?
- how much interaction can other participants (councillors or members of Parliament, moderating or management staff of the agency) *realistically* commit to?

As always, the answer often reflects a compromise between these competing forces.

---

[6] The online discussion forum previously run at Moreland City Council in Melbourne was specifically implemented by a Councillor for this purpose.

> **Exhibit 17: SMS Voting – Australia versus the United Kingdom**
>
> In the recent extensive pilot of remote voting systems in the United Kingdom, SMS voting was used as one of a number of channels for the voting process (internet, electronic voting booths, paper ballots, postal, mobile telephones). In this example, SMS was seen as particularly valuable in reaching the youth demographic, given a good fit between this channel and that audience. While this was valuable in the UK, the use of compulsory preferential voting systems (and the need to be able to decipher a long ballot paper and navigate it easily) employed in Australia makes this channel of little value.

3.1.6. Choosing the Right Channel(s)

Following consideration of the issues raised above, the selection of the most appropriate channel moves (hopefully) from being a complex technical assessment process, to one where the manager can discriminate between differing engagement needs and balance these against different characteristics (strengths) of various technical options.

The core questions during this phase of the planning process are:

- what are the available technical options?
- what are their characteristics?
- how do these match the audience needs?
- how do these reflect the objectives of the project?
- to what extent do they afford the degree and type of interactivity required?

Different target audiences will have higher levels of familiarity with different communications channels than others. A good illustration of this is the way different generations use technologies, like mobile telephones. While older people generally treat these as portable analogues of the landline telephones they grew up with (predominately using voice), younger people are more likely (for a range of reasons, including cost) to demonstrate a higher ratio of SMS to voice using the same technology.

While any generalisations must be treated sceptically, there is a general tendency for people to develop *information literacy*,[7] technical expertise and confidence in communications technologies (and specific submedia or applications[8]) and

---

[7] Defined by the US National Forum on Information Literacy as 'the ability to know when there is a need for information, to be able to identify, locate, evaluate and effectively use that information for the issue or problem at hand'.

[8] Here it is important to separate technologies and applications. Example 1: a technology may be a telephone, but have a number of applications (one-to-one voice calls, conference calling, chat lines). Example 2: the Internet is primarily a specific form of networking, but has a wide array of applications (submedia) that often share little functional similarity (web browsing, e-mail, P2P, online gaming, IRC,

software during particular stages of their lifecycle: as they reach adulthood, during their years of formal education and/or the onset of professional employment. This may explain why your parents enjoy media you find uninteresting, while your children enjoy media you find incomprehensible.

Exhibit 18: SMS Consultation

Lancashire County Council uses SMS messages to prompt members of the community for quick responses to a range of questions under consideration. The popularity of SMS, particularly among younger people, provides the Council with the opportunities to target particular elements of the wider community. As the communication channel employed has specific characteristics that limit verbose participation (tending to generate fast feedback, but with short responses), it is useful for a specific type of engagement process (educational and idea generation), but is unsuited for more complex policy issues. See: http://www.lancashire.gov.uk/environment/sms/index.asp

---

The important point to remember is that selecting the right communications channel can significantly influence the type of participants likely to be attracted to the process. This can sometimes be beneficial, especially if the aim is to achieve segmentation (such as the use of chat for younger people), however, if the aim is to capture a wide cross-section of the community:

- broad-brush applications are preferable (email, web-based participation, mobile telephony); or
- multiple applications (channels) will need to be employed.

One of the ongoing strengths of ICTs is the growing convergence of different types of communication applications upon a standard internet protocol that is becoming increasingly available across a wide array of devices.

We can expect to see decreasing costs associated with the incorporation of multiple channels within a single system, or approach, over coming years – making broad-based strategies more cost-effective. The trend towards convergence will also be valuable in reducing administrative costs (for example, seeing web-based interaction and mobile telephone participation served by the same database and management system).

3.2. Concept Development Approach

Depending upon the scope and scale of the issue, the tractability of the participatory problem and the importance of the policy development process,

---

etc.). Thus, a user may be regarded as technically proficient in one application of a technology, but not another.

the establishment of a project team and development of an eEngagement approach can occur quickly, or represent a stand-alone consultative process in its own right.

Exhibit 19: 'Full Service' Commercial eEngagement Providers

A number of private firms have begun to offer electronic research and consultation services, including the capacity to provide 'full service' provision of electronic consultations from conceptualisation through to implementation.

These providers can be useful where a public organisation may:

- lack the skills needed to undertake planning and implementation;
- be only interested in a one-off process and uninterested in developing organisational skills and infrastructure;
- look to partner with existing providers in the early days of developing eEngagement capabilities to increase their speed of learning; and/or
- be interested in a specific technology provided by a private firm.

Some examples of these types of providers would include:

- Insightrix, an online research and consultation service provider [http://www.insightrix.com/];
- Ezicomms, a provider of handheld devices which allow for interactive 'town hall' meetings [http://www.ezicomms.com/];
- BigPulse, a provider of 'online opinion markets' [http://www.bigpulse.com/];
- Securevote, a specialist provider of secure online voting systems [http://www.securevote.com.au/];
- Everyone Counts, a company that develops and provides online surveys, polls and elections [http://www.everyonecounts.com/];
- National Forum, a non-profit organisation with experience in developing interactive websites for government and political organisations [http://portal.nationalforum.com.au/]; and
- Social Change Online, another non-profit organisation that develops web-based services and provides moderation staff for public enterprises [http://online.socialchange.net.au/].

The combination of: (a) a relatively simple or straightforward policy issue, and (b) a clear fit between a target audience and particular channel, may encourage a relatively rapid concept development process. However, the establishment of any eEngagement process requires lateral and creative thinking in order to anticipate and consider the range of alternative approaches. The tendency for some online consultation and democracy projects to overemphasise available tool sets can sometimes lead to a process driven by available technologies, at the expense of approaches that might yield a better outcome.

Examples of concept development approaches (from the simple to the highly complex) include:

- in-house development only;
- co-operative public sector development (agency plus intra-governmental stakeholders);
- *ad hoc* community consultation (liaison, often with loose timeframes; 'ring around');
- Request-for-Information (RFI) consultation (formal, submission timeframes and loose expected format);
- formal consultative approach (release of documentation, *pro forma* submission templates, etc.);
- working / planning / brainstorming day(s);
- workshops and road shows;
- use of an external research service; and
- a combination of the above approaches (as illustrated in Figure 8).

Figure 8: Complex Concept Development Process (Australian Tax Office)[a]

Issue	Discover	Invent	Evaluate	Survey
'How does the organisation respond to the community?' (benchmarking current performance)	User clinics to understand issues (focus group approach)	Creative retreats (intensive brain-storming and idea generation)	User observation tests of prototypes	Post implementation survey
Time →				

[a] Adapted from: Vivian, Raelene 2004, *Elements of Good Government Community Collaboration*, Discussion paper no. 2, Australian Government Information Management Office, Canberra, <http://www.agimo.gov.au/publications/2004/05/egovt_challenges/community/collaboration>

3.3. Managing Identity Issues

One issue that all eEngagement processes face is that of the identity of participants in the process. While it is often difficult (or unnecessary) to determine the identity of individuals in *physical* meetings, the capacity for people to participate using ICTs from any location gives rise to public sector managers' concerns about the potential for misrepresentation in the eEngagement process. Depending on the process being developed, this may be a significant issue (e.g. for an electronic voting system), or of little or no consequence (such as when collecting *ad hoc* comments online for a minor issue).

At the level of technical management, however, managing identity in online participation can be one of the most complex and difficult areas of decision-making associated with developing eEngagement approaches. The issue of identity has two dimensions:

- desirability and notions of eligibility; and
- technical aspects of identification (proof).

3.3.1. Desirability of Identification

The desirability of identity management for online systems is the first question that needs to be addressed in development of any approach to managing personal information in the online consultative, or participative process. The primary question that needs to be addressed here is, whether there is some basis for *exclusion* from the participatory process.

This may appear, on the surface, to reflect a negative approach, however, the question is underpinned by the following considerations:

- is there some legal, socio-cultural, or moral restriction to be placed on participation and why? This may include examples where:
    - the issue relates to a local area, with implications restricted to that area alone;
    - participation is a right of citizenship only;
    - there are concerns about age of consent issues for participation;
    - the issue concerns current recipients of a service;
    - the eEngagement process has been designed to specifically counteract under-representation of a minority group;
- there is a practical reason associated with the restriction (such as limiting participation numbers). This may be the case where the issue is popular and would attract a large number of non-affected 'hangers on';
- the audience has been specifically selected to adhere to a particular mix of characteristics (e.g. quota sampling) and free access to participation would undermine this approach; and/or
- the issue is particularly sensitive and is being carried out in a highly controlled and managed environment.

While it may appear obvious that entry into the process will be controlled, it is not always clear that restrictions on participation need to be enforced. Exclusion from a participation process can be difficult to justify to affected persons or stakeholder groups, particularly if:

- participation is restricted, but the planners failed to identify a relevant stakeholder group prior to the 'rules' being developed;
- the process is not binding in nature; and/or
- if benefits appear to accrue to persons or groups participating in the process (such as social connectedness) which are denied to others.

Overall, the question of eligibility can be broken into three levels:

- *no verification is necessary* (least common) – participation is open to all;
- *some verification is desirable* (most common) – casual or troublesome participants are discouraged by a formal registration process (self-completion); and
- *absolute verification is required* (uncommon) – the participants are specifically identified against some form of independent, or absolute, system of identification which contains their relevant proof of eligibility (e.g. electoral role, drivers' licence, etc.).

Exhibit 20: Is This a Local Issue?

In the development of a citizen-based consultative process to develop alternatives to the official World Trade Centre re-development process, the *America Speaks* project team limited participation in the online forum to people living in and around New York. The team soon received requests to participate from across the United States and while these requests were politely declined, people from outside of New York managed to find their way into the process.

When asked, these people stated that the World Trade Centre was an American issue, not one simply for residents of New York and that they had strong personal feelings about how the site was being treated following the 9/11 attacks. They implicitly questioned the eEngagement managers notions of who had 'legitimacy' to participate in debate surrounding the redevelopment of what would be an iconic national project.

They were allowed to take part.

One of the important issues to remember in this early phase of decision-making, is how restrictions on participation (or the lack thereof) shape outcomes. In some circumstances, it may be considered necessary to apply *controls* whilst not discouraging broader participation. In such cases, a two-step process may be required that allows open participation in more 'general' forums, on the one hand (such as participation in a discussion forum), with restricted participation in *deliberative* forums (targeting individuals or groups falling under a specific category [citizens], or though a secondary sampling system, such as delegation to a group of elected spokespersons).

3.3.2. Technical Aspects of Identification

Following the determination of the necessary levels of eligibility and identity verification, the next question (and one which will shape the technology used for the consultative process), will pertain to the technical means by which identification can be assessed (either to manage access to the system, or as part of the pre-participation approval process).

Exhibit 21: Using 'Cookies'

While some computers have a fixed internet address which allows websites to identify them on an ongoing basis, most computers do not, making it difficult to identify a user from one visit to the next. To manage this difficulty many websites use 'cookies'. A cookie is a small computer file placed on a user's hard drive to record data about a previous visits to the website or service. The cookie allows a computer to be identified and information stored about that computer's activities. Cookies can be useful in:

- storing preferences about how webpages should be displayed;
- storing user identification information to allow the user to 'automatically' log into a web service;
- retain a memory of the user's activities or pages visited; and
- developing a usage pattern for users to improve the service or information structure.

While these advantages are significant, there are also problems associated with this approach:

- some users will not accept or use cookies, either because of concerns about privacy, because the computer they use is unable to accept them, or because they use a shared computer (such as a public access terminal);
- while some users secure their computer by using a personal password, not all do – authenticating via cookie only authenticates the *computer*, not the person using it. Allowing a cookie to automatically authenticate a user may allow a third party to impersonate the user; and
- cookies can identify websites that have been visited by the user, this may be undesirable if the issue is sensitive or the user is at risk (e.g. a consultation associated with family violence).

These technical questions are best undertaken in direct consultation with security and IT staff and must, at least, include consideration of:

- the existing infrastructure surrounding identity in your agency (and the distribution of tokens, passwords, or similar systems to potential participants);
- existing authentication technologies (e.g. public key infrastructure); and
- the necessity to develop technical separation from token to identity.

This last point is particularly relevant where the administering agency uses pre-existing information about the participant and uses this to collect policy-related or personal information. In this case, privacy legislation will require physical or electronic separation between the corporate knowledge used to provide secure entry into a system and the information provided by individuals during the eEngagement process.

In addition, careful consideration of this approach will be necessary where the eEngagement process requires both user *validation* and user *anonymity*. This can mean either complete anonymity throughout the whole process, or levels thereof – such as anonymity within a discussion forum (between members of the public), but where the agency has the capacity to identify and follow-up on specific participants.

Overall, online identity management – to greater or lesser degrees – depends on issues that are outside the control of the agency (such as the ability of the users to ensure that they have a secure computing environment, or their capacity to remember and keep passwords secret, etc.).

While an agency may develop a robust security and identify verification approach, this can be undermined by users themselves. Security and identity supervision is about risk management and reducing the *probability* associated with fraud or impersonation.

4. Implementation

Having developed an approach for the eEngagement, the implementation phase represents the realisation of this vision. Often, this will require stakeholder input – across government and outside of it – and necessary adjustments to the initial plan in the light of unforseen eventualities. In this way, implementation is just like any other process for project delivery.

Rather than provide a summary of issues associated with standard project implementation and management issues, this section (and the later discussion of post-implementation issues) focuses on aspects of specific, or particular, relevance to the manager engaged in eEngagement activities.

4.1. Stakeholder Buy-in

The first step in successful realisation of the eEngagement approach is ensuring appropriate commitment from key stakeholders. This may entail a new process of negotiation, 'selling' and discussion, or may reflect the formalisation of processes already undertaken as part of the visioning process.

Four important considerations are:

- *managing upwards* by ensuring appropriate commitment from senior policy makers (managerial or Executive). This will be of particular importance where the process has a deliberative element, where for all intents and purposes, the project team is asking for the engagement process to be *delegative* in character;
- *managing sideways* by intra- and inter- governmental stakeholders may need considerable persuasion, either to establish their commitment to the process, or to provide resources and participation within it, or because of the need for their (possibly long) approval and authorisation processes to be undertaken;
- *managing outwards* by identifying and ensuring commitment from members of the community to the process and approach to be taken;
- *managing inwards* by ensuring that staff are committed to the process and the approach. Internal resistance can be one of the least-well-managed factors in the successful realisation of the plan.

4.2. Developing an Engagement Plan

Once stakeholder commitment is established, a formal engagement plan will be developed. This takes the form of a formal work plan and might simply be an elaboration of the proposals and associated documentation developed to date. The work plan needs to address:

- formal timeframe commitments;

- budgetary allocation;
- participant role descriptions;
- output indicators;
- evaluation approaches and success measures; and
- contingency strategies.

In addition, the engagement plan can serve as a formal, or informal 'contract' with the public. In areas where participation has been poor because of low levels of trust, making this document participative, or public, can be useful to demonstrate commitment to the engagement approach by the agency and provide a benchmark against which agency performance can be observed by stakeholders and potential participants.

As the implementation process moves forward, the formal engagement plan can serve as the basis for supporting documentation such as:

- the marketing and promotion strategy;
- the technical specifications and, if necessary, contracting documents for systems development;
- evaluation frameworks;
- the final report; and
- the evaluation report.

Good documentation, from the outset of the project, will greatly assist in the process of post-project review and project termination.

4.3. Managing Technical Implementation

Managing the technical aspects of the project implementation process can be the more complex area of the implementation plan, particularly for managers who are not highly familiar with the technologies supporting the engagement strategy.

There is a risk that lack of technical familiarity or knowledge might, perversely, lead managers to 'outsource' the technical side of the process to a private company or IT unit in government. It must, therefore, be emphasised, that maintaining strong control and oversight of this part of the process will be critical in ensuring that the objectives of the eEngagement process are realised.

Although some IT professionals have a very good grasp of the social issues arising at the interface of computer technology and public policy, many have not and the collaboration between policy specialists and technologists can be the most productive and educative part of the implementation process. In many cases, existing relationships and associated business processes governing the provision of IT services to the organisation may require some (re)negotiation (either to determine the ability of the existing provider to undertake this work, or to allow new systems to interact with existing ones).

Technical implementation will require:

- consideration of the relationship of this project with existing agency or departmental IT strategic plans;
- acquisition of appropriate expertise or advice (either through the appointment of a technical staff member, or through an out-sourcing arrangement);
- assessment of technology options and vetting against eEngagement objectives;
- determining an appropriate 'solution' (choosing the product or bundle of services and products);
- managing costs and implementation timeframes;
- the 'purchase' or commitment decision; and
- management of implementation, user testing, review and 'going live'.

Exhibit 22: City of Wellington IT Strategy

In their 2006 IT Strategic Policy Document, the City of Wellington in New Zealand has incorporated electronic democracy as one of the three elements of their IT approach.

The eDemocracy element of the strategy includes four objectives:

- accessible information;
- accessible elected members and Council officers;
- encouraging broader consultation; and
- efficient services.

The document includes a discussion of the implementation approach and an explicit identification of the need for post implementation assessment of the approach.

The policy can be located at: http://www.wellington.govt.nz/plans/policies/ict/pdfs/ictpolicy.pdf

---

4.3.1. Determining the Software Feature Set

The *software feature set* is the bundle of attributes possessed by the application that achieves the objectives of the engagement process. This is normally constructed as a list of 'can dos' – the software can do x, y and z to meet the objectives. Depending on the budget of the organisation and the complexity of the eEngagement process, this may need to be clustered into 'must have', 'should have' and 'would like to have' characteristics, thus supporting an analytical approach to making trade-off decisions (where necessary).

This approach may simply require the development of a list, or may be expressed as an analogy, such as 'the system should emulate a library, with a check out area, reference table, volumes of texts, etc.' Do not underestimate the value of drawing analogies: they can generate powerful metaphors that assist in visioning and ease communication between purchasers and providers (and later may be valuable in assisting users to understand the parameters of the interactive environment).

It will be useful to engage the participation of a technical advisor early-on in the process, as some activities that appear difficult or complex to those unfamiliar with the technology may in fact be commonplace and straightforward, or vice-versa. Having early advice can also serve to generate new ideas prior to systems development which the project team may find valuable. This could take the form of new features, or potential features, that will be noted for future iterations of the approach or as contingencies.

4.3.2. Who Governs? Technical, Administrative, or Political

Regardless of the approach undertaken, a critical decision point is the determination of who is ultimately responsible for developing the technical package. For various reasons, the implementation process may require, inter alia: a strict level of control by the auspicing agency; IT or communications technology experts having autonomy with respect to decisions about technical issues; or direct hands-on management by a Minister or select committee (e.g. in a Parliamentary process).

Decisions about software acquisition, management and/or modification may entail an explicit choice between close management of the process by the host agency, or devolution of the process to a technical unit, or private firm. It is important to be cognisant of the advantages and limitations of each approach:

Figure 9: Who Governs?

	Advantages	Limitations
Technical Experts	Strong technical understanding, solid project management skills, awareness of advantages and limitations of subtle differences in technical design, understanding interoperability issues (particularly for data interchange).	Can over-focus on technology, can emphasise effective technical management at expense of engagement objectives. Lack of policy and engagement expertise.
Administrative / Policy Staff	Strong understanding of policy-making processes, the issue(s) under consideration and the wider environment (stakeholders).	Sometimes very weak in technical understanding and insensitive to technical concerns and timeframes.
Political Layer	Ability to commit resources and devolve decision-making, clear articulation of importance of project through 'demonstrated project leadership'.	Often unable to commit time to project, can over-focus on political benefits and short term electoral cycle issues.

One potentially valuable approach to is to place project management *outside* of government. This may take the form of outsourcing a specific aspect of the project (such as the development and maintenance of the technical infrastructure), or shifting the entire project to an organisation within (or with explicit expertise in) the community of interest, such as an academic body, or a non-government organisation (NGO).

The advantages of this approach include:

- external organisations may have expertise not possessed by the host agency;
- third parties can serve as a 'buffer' or neutral arbiter;
- low cost facilitation of technology exchange between principle and agent;
- eEngagement process can be 'insulated' in the event of failure;
- use of non-profit or academic agencies can be cost effective;
- fosters community of expertise, or marketplace;
- access to new networks (social, professional); and/or
- the capacity to capture additional, or 'multiplying', sources of funding (such as integration of research funding).

The limitations of this approach include:

- loss of control over process (to a greater or lesser degree);
- additional layers of negotiation and project management;
- transfer of skills may 'de-skill' organisation; and/or
- it is sometimes difficult to identify third party organisations having the necessary skill sets and capacity.

Exhibit 23: Placing Management of the Participatory Process Outside of Government

In the case of the Hansard Society's collaboration with the Social Security Select Committee in the United Kingdom's *Uspeak* online consultation project, the online consultation process was undertaken by a collaborative *umbrella management* structure including voluntary organisations with expertise in the specific policy area and target audience, local government organisations capable of providing place-based assistance with recruitment and promotion and private sector providers with capabilities in developing access technologies.

The Department of Human Services (Victoria) uses an external, non-profit organisation to provide moderation skills for their online discussion forums. In this case, the Department benefits from acquiring the necessary skills without long lead-times associated with recruitment and training.

4.3.3. Make or Buy?

The most fundamental decision in the development of an eEngagement technology platform is choosing the right type of software to meet the project requirements (as discussed in Section 3, Designing the Right Approach). Depending on the specific engagement strategy being undertaken there may be a range of software packages available to host and administer the online engagement activity.

For eEngagement projects that are based on electronic discussion list models, there is a range of existing software packages that allow for these types of discussions to be hosted, either as simple email handling systems, real-time chat facilities, or Web-based bulletin board systems. For more complex or innovative projects, the lack of a large commercial marketplace for electronic democracy software means that there may be few, if any, off-the-shelf software packages available.

A critical early decision will address whether existing software packages can deliver the functionality required for the proposed engagement activity. This will also require consideration of:

- the degree to which the software packages interface with other systems to be employed by the project management team, such as database systems for handling contacts (e.g. the corporate Customer Relationship Management system) or managing project timelines and data analysis tools to assist in effective evaluation of materials collected;
- the capacity of existing information technology infrastructure to run or host the software under consideration (e.g. a bulletin board system built using a computer language like PHP may require a posting server to have specific capabilities, such as a particular database product); and
- the capacity to customise the software being considered, if required, to meet the needs of the consultation process. In addition, a highly flexible (e.g. feature rich) or customisable software package may be required if the engagement process is intended to accommodate significant input by the participants in shaping the process and/or provide an interactive environment or decision-making process.

These considerations may in turn require further decisions about whether to select an existing software package (which may necessitate a trade-off between availability versus functionality), the re-engineering of existing applications, or the development of wholly new applications to undertake the task.

Even when undertaking a relatively conventional approach – e.g. in which an existing software package is available to the project team – the likelihood exists that some form of customisation or modification will be required to accommodate the agency's requirements, such as:

- integrating the system with appropriate document management protocols and storage systems (such as systems to retain documents for FOI or archival purposes);
- integration of the system with existing IT security systems, such as user login profiles or workflow management software (e.g. groupware or Lotus Notes); and/or
- data interchange systems to allow the importation or exportation of data from one database to another (e.g. moving textual information from a discussion list into an analysis package, such as NVIVO).

4.3.3.1. Do we Need New Tools at All?

Public sector managers are often surprised to learn that their agency (and partner organisations) either possess, or can access under licence, a wide array of applications software capable of supporting eEngagement activities. As part of any assessment of available technologies, it is important to consult an inventory of available software (or undertake an inventory, if none exists) and assess the utility of these packages in meeting the objectives of the eEngagement plan.

This is a useful strategy where resources are limited, or the eEngagement process is not technically complex – it does *not* imply a process in which the objectives are retrofitted to the tools on hand. Public agencies commonly possess, or have access to technologies like:

- email management systems;
- servers, with scripting capabilities and database integration (often with good security);
- content management systems (website management engines), which may include:
    - simple polling and survey design and management systems;
    - online discussion facilities (bulletin board systems);
    - email collection systems;
    - password access (password-restricted access); and/or
    - groupware systems (such as intranet systems). For example, the 'Central Station' intranet system of the Victorian Public Service includes the capacity to develop 'communities'. These communities allow for online discussion, online publication and the lodgement of shared documents.

In addition, the desktop environment of many public agencies also have an array of potentially useful software, such as:

- statistical analysis packages (such as SPSS or R);
- qualitative analysis packages (e.g. NUD*IST/N6);
- database and spreadsheet packages (e.g. Access, FileMaker Pro, Excel); or
- desktop publishing applications.

Alternatively, non-government 'communities' tools abound, some of which may be of value, such as:

- Yahoo!, MSN, or VicNET communities;
- blogging and publication tools;
- online meetings sites; and
- commercial, low-cost, online surveying packages.

4.3.3.2. Purchase Point Considerations

The decision about whether to 'buy or build' software is a complex undertaking and will need to be considered with due reference to the existing agency or whole-of-government policy on software acquisition.[1]

The decision about whether to purchase an existing software package, or to develop a custom-made piece of software will be based on:

- a balanced assessment of the trade-off between cost and functionality. A decision to 'buy' can be supported on 'value for money' grounds, where there is a high degree of *fit* between the functionality offered by existing software packages and the requirements of the eEngagement plan. Purchasing an existing software product can represent significant time savings over the lifecycle of the project. Depending on the vendor, a collaborative approach to acquisition might be considered in order to improve the alignment between the existing functionality and project needs. This also offers significant advantages to the agency and the vendor by allowing the agency to develop a better software system for its specific needs, with costs shifted to the vendor, while allowing the vendor to develop their product in the expectation of improving its marketability in the future; and
- the practical capacity of the agency, in collaboration with supporting information technology units, to undertake an in-house software development process. While there is considerable expertise in developing software within the public sector, this expertise is often rationed towards core corporate information technology systems over 'line' applications. If the agency feels that it lacks the expertise to manage a software development process, this will be a mitigating factor against significant internal development.

Regardless of the approach taken, a clear business case process will be undertaken in line with existing policy in the agency's jurisdiction.

---

[1] For example, *A Guide to ICT Sourcing for Australian Government Agencies* (Australian Commonwealth Government) or *New Zealand Government Information Systems Policies and Standards*. Similar documents and policy statements are maintained by each State and Territory in Australia.

4.3.3.3. Proprietary versus Open Source

When considering whether to 'make or buy', public sector agencies also consider non-commercial software alternatives. During the last 10 years there has been an increasing interest from the public sector in 'open source software'.

Open source software is provided under a *licence* that commonly allows for its original source code to be freely distributed and modified by end users. While open source software is often considered to be 'public domain' intellectual property,[2] this is not the case. Open source software is often released under a licence that imposes specific restrictions on the end user or modifier. These may include:

- requirements to preserve the identity of contributing programmers and developers;
- restrictions on the use of the code for commercial purposes (e.g. reselling);
- requirements for development of the initial code to be released under a similar licensing arrangement; and/or
- exclusions and indemnities.

The growth of the 'open source movement' has stimulated the proliferation of software packages that are available free, or at low cost to the public, as well as encouraging the release of software which may have been developed for specific purposes but does not have a commercial value. As this movement has developed into a strong, collaborative community, the availability of elements of code and whole applications allows new software applications to be developed from existing code, without undertaking complete software development – in other words, elements from one project can be incorporated into another to add functionality.

The advantages of utilising open source software are:

- low-cost for the initial acquisition of the software package. Open source software can often be downloaded for free, or simply for the cost of the media and supporting documentation. Some open source packages (such as the popular replacement for Microsoft Office – OpenOffice.org) have been developed with the explicit objective of high usability, making installation and operation no more difficult than commercial offerings;
- access to a large community of developers who can assist and advise in the development of the software package;
- the software can be modified simply because the source code is open for modification (whereas proprietary software often prevents modification

---

[2] Intellectual property that has no proprietary claims made against it (e.g. no ownership). A good source for information on open source licensing is the Free Software Foundation (http://www.fsf.org) or the Open Source Initiative (http://www.opensource.org/).

outside of the vendor's company or requires the purchase of developer's kits);
- the contribution by government of new applications can grow the user and developer base, increasing the number of people working on the software;
- proprietary software can become orphaned (abandoned completely or no longer offering upgrades or support) which necessitates shifting to a new product; and
- modified versions of the software can be developed and distributed freely without reference to any vendor.

The disadvantages of utilising open source software are:

- lacking vendor support, open source projects require either the participation of a motivated volunteer programmer community, or the investment of time and money in developing and modifying the software in-house;
- some open source projects are poorly documented, making modification difficult;
- open source software can lack the 'polish' of commercial software, particularly in terms of user documentation;
- open source projects can become inactive leading to limited further development;
- the software may be at an early stage of development, leading to a quick succession of releases which necessitates regular updating; and
- many government information technology agencies (central policy and standards bodies) are cautious about the use of open source, because of concerns about malicious code, or the poor quality of software employed.

Open source software can deliver some cost savings to government agencies interested in developing unique packages. However, it is a misconception to consider open source as 'free' software. Depending on the type of functionality required, considerable investment may be necessary to develop or modify an existing open source application. In addition, care must be taking when adopting an open source solution, given the variety of licences that may be attached to the original code and the constraints these might place on further development or release of modified versions of the initial software.

The real advantage of open source lies in the ability to redistribute modified versions of the software to organisations with whom your agency may be partnered. For example, you may develop an online consultation system in open source that is of interest to a number of peak industry bodies engaged in the initial consultation, who wish to use the software to consult with their members. Having developed the package in open source may allow your agency to freely release its software to partners/stakeholders and any enhancements that they make to the software can be acquired by your agency for future use.[3]

4.3.4. Low Tech versus High Tech

A common but largely avoidable problem associated with the design of electronic democracy systems arises from the mistaken belief that, because these concepts are new, they will be supported by the latest technology. For example, the capacity of computers and mobile telephones to support interactive multimedia often leads to their being selected to deliver eEngagement solutions. This assumption is often ill-founded and can limit the degree of participation by members of the public.

Over the last decade, the public sector in Australasia has become much more effective at managing issues of technological obsolescence, through systematic hardware and software replacement processes and the use of managed service agreements (equipment licensing). While this has positive benefits in terms of the productivity of public servants, public sector managers need to be cautious when assuming that members of the public have similar technical capacity.

This has a number of dimensions:

- on average, the public sector has newer hardware and software than the general community. The connection speed at which the public access the internet and the capacity of their installed software is much more variable than that found in the public sector;
- public employees on average have higher levels of information literacy than the public. As 'white collar' workers, public employees have greater levels of experience using IT systems than the general community; and
- public sector employees are disproportionately *urban*. Urban areas commonly enjoy higher connection speeds and greater reliability of telecommunications services, which in turn leads to higher internet usage and like services (bandwidth).

Technological solutions need to be matched to the target audience taking into account its general characteristics, like technical skills, technological capabilities and information literacy. Target audiences can be grouped according to a number of archetypical user types (as illustrated in Figure 10).

---

[3] The Australian Federal government has developed a guide for public sector agencies interested in acquiring open-source software. This guide can be found at: http://www.agimo.gov.au/_sourceit/sourceit. The New Zealand National Government has also given consideration to some of the issues associated with open source software, a discussion of which is located at: http://www.e.govt.nz/

Figure 10: Archetypical Internet User Types

'Experimenter'	'Sporadic User'	'General User'	'Power User'
• Low levels of use • Simple applications (web) • Exploring and learning • 'Lingers' on websites • May only use technology at a location remote from the home (work, internet café, public access terminal)	• Infrequent, irregular use (borrowed or public access terminal) • Sees value only in occasional use (specific purpose)/or • Access may be limited by situation (access, affordability) • May have a internet connection in the home	• Regular use for short periods of time • Sticks to tried and true software ('standard desktop software') • Emails existing social / professional network • Consumes content • 'Trusted' transactions (banking, taxation) • Likely to retain a slow internet connection	• High levels of frequent use, invests in technology • Experiments with new technology / software • Creates content • Develops new online networks • Many transactions online • Likely to invest in high-speed internet access

A common trap for public sector managers is to assume parity between the capacities of government officials and those of the public with whom they engage. Examples include:

- distributing unnecessarily large documents online, or designing websites that include extraneous formatting or images that are slow to access for users with older hardware or slower connection speeds;
- using file formats or website elements that require the most recent software to access, thus requiring the end-user to install software that is not commonly used by most web users;
- requiring a platform- or technology-dependent capability (such as a specific type of internet browser or screen resolution); and/or
- providing few choices to the user as to how they interact with the system (such as web-based electronic discussion lists that do not permit users to access the discussion through a generic mail client, such as Outlook or Eudora).

When selecting the most appropriate approach to technology, consider the following four questions:

- do we need this feature to achieve our objective (is it simply aesthetic, 'cool', or redundant)?
- does the system offer the end-user a choice in the range of technologies used to access the system (is it technologically *agnostic*)?
- does the system work with older technology and slower access speeds?
- is the means of accessing the system part of the standard operating environment of most computers/ICTs?

If the answers indicate that the user base is most likely to be served by older or more common technology then work within those limitations. While this may restrict some activities, it does not necessarily prevent the eEngagement process being interactive and compelling. Having users with slower and older software can allow interactivity and complexity of systems design, but may require:

- multiple approaches to displaying the content, such as inclusion of a 'text only' version (see section 4.4.1, Compelling Content versus Eyecandy); and/or
- placing the bulk of information processing at the server (agency hardware) end of the communications channel (e.g. putting the 'smart' end of the system within the computing environment of the agency).

4.4. Generating Compelling Content

At the end of the day, *content is king*. The information you provide, the quality of debate you generate and interactivity can be essential in shifting participation and interest from passive to active. Content is not, however, about graphic design. Although style is important for some target audiences, the quality of the information presented, including its clarity and accessibility, can encourage or discourage participation.

Public sector managers need to understand that we are all living in an increasingly media-rich environment. Although the consultative processes of public sector organisations are by their very nature important, they do have to compete with a wide range of demands on the time and attention of potential participants. In some cases people consume multiple media simultaneously and each source of information must compete for attention immediately. It is essential to recognise that many target audiences are continually confronted with the need to ration or make 'tradeoffs' in their use of communications technologies. Therefore, when attempting to encourage participation in eEngagement initiatives, the core challenge of content development is to reward those tradeoffs with compelling content.

Some suggestions for encouraging participation in eEngagement initiatives include:

- use plain language;
- use short summaries that allow participation based on 'skimming' (provide extensive content as options for further reading);
- provide alternative interpretations of information (such as 'case studies', points of view, or first-person accounts);
- allow issues to be personalised (e.g. if the issue is one involving cost tradeoffs, provide an online calculator to determine benefits at different levels of costs for the individual);
- where appropriate provide a mix of media forms (text, audio, animation, diagrams, games) which allow users to consume the content in their preferred format;
- workshop, or user-test, content with a focus group (formal or informal) prior to release;
- be prepared to adjust content 'on the fly';
- provide capacity for participants to generate their own content;

Electronic Engagement

- ensure an appropriate 'refresh' period for content – eEngagement processes that attract commitment provide new information and experience on each visit by the participant (where the process is multi-stage); and
- remember what the user has seen and make navigation menus dynamic to present new or unread information on the next visit (requires cookies).

Exhibit 24: Public Participation Geographic Information Systems

Geographic Information Systems (GIS) is a type of mapping software that allows the storage, analysis and presentation of spatial data. GIS systems are used largely in land / urban planning processes, but allow data to be overlayed for analytical purposes. Data may include land use, pollution flows, car movements or any other form of information pertaining to location and position.

These systems can be very useful for relevant eEngagement purposes (Public Participation GIS (PPGIS)), allowing participants to visualise and analyse spatial issues, or provide data to overlay existing map data. With global positioning (GPS) being incorporated within some low-cost consumer electronics, the capacity for members of the community to contribute to GIS datasets (rather than simply consume data) will expand. Examples of PPGIS include:

- as part of a major review of their Local Environment Plan the Kiama Municipal Council partnered with the University of Wollongong to develop a web-based GIS site to allow members of the community to visualise land use issues in the municipality. The system allows members of the community to look at current land use issues across the whole municipality before completing a survey;
- the Community Block Grant Administration of Milwaukee has employed GIS in local neighbourhood strategic planning, where members of local communities undertake assessments of local strengths and needs based on data provided on economic and social indicators and presented using special mapping techniques;
- GIS has been combined with 3D imaging technology to allow for the 'visualisation' of different policy decisions for land use and area planning, allowing communities to see the projected impacts of different land planning regimes on local growth and the aesthetics of the community. See: http://www.communityviz.com/

4.4.1. Compelling Content versus Eyecandy[4]

The multimedia elements of ICTs are often touted as one of their compelling features. The convergence of text, audio and video can allow the development of attractive and entertaining online content that presents – often dry – content in a dynamic manner. When developing online content, it is important to gauge the value of dynamic media against technical issues (discussed in Section 4.3.4, Low Tech versus High Tech), as well as the relevance of stylistic design to the objectives of the engagement process.

Exhibit 25: Web Design for Accessibility

When utilising a website as a primary or secondary element for eEngagement, it is important to apply the relevant World Wide Web Consortium (W3C) guidelines to ensure that the content is presented in a manner accessible (readable or interpretable) to the widest possible audience.

These standards have been put into place to assist website designers to ensure that content is accessible to users who may:

- have limited vision or dexterity;
- have learning impairments or poor language skills;
- use assistive technologies to encounter information online (such as text-to-speech converters or Braille computers); and
- have low information literacy skills.

The guidelines provide technical *and* stylistic suggestions to increase the readability of online content and are mandated by many levels of government in Australasia. While generally considered a requirement for disabled members of the community, these standards have wider value to people whose primary language is not English, older members of the community and people who have poor literacy. Because the Australian and New Zealand societies are progressively aging, consistent application of these design guides will be increasingly important for social inclusion.

Further information on the guidelines can be found at: http://www.w3.org/WAI/

---

[4] Eyecandy (n) is defined by the Labor Law dictionary as 'visual images that are pleasing to see but are intellectually undemanding'.

The tendency to over-emphasise design can significantly limit participation by:

- misrepresenting the content as not serious;
- preventing access by people with limited access speeds or who are unfamiliar with complex graphical user environments; and
- slow interaction with the system.

The accessibility of ICT interfaces is an ongoing concern for all governments for a number of reasons:

- governments are keen to promote their online information as accessible to all;
- there have been cases of litigation where inappropriate site design has prevented participation (for example: *Maguire v SOCOG*);
- the convergence of the internet protocol with a range of electronic devices makes 'standardised' designs (such as websites that enforce a specific screen resolution) unwise; and
- there is a growing movement towards standardisation of presentation to aid consistent branding and make user navigation simpler.

A good web designer is invaluable in ensuring that these issues are well-managed. They will have:

- a good technical understanding;
- a good understanding of useability issues;
- an awareness of relevant standards bodies and guides; and
- an awareness of relevant legal requirements and risks.

4.5. Promotion and Recruitment

Promotion and recruitment is one of the key requirements for the development of a successful eEngagement (or any other consultative) project. One of the primary tests of eEngagement and online consultation activities is the extent to which the process has attracted participation. As the decline of civic participation (see Exhibit 1) is commonly the core motivation for government interest in eDemocracy activities, the success or otherwise of promotion and recruitment (and later retention, see Section 5.2, Closeout Processes) will often come to define the success of the activity in the mind of senior managers and Ministers.

The appropriate approach to promote the eEngagement process and recruit participants will depend on the nature of the process being undertaken, its objectives (particularly expectations of large or small numbers of participants) and the characteristics of the target audience.

Most government agencies use a combination of:

- website placement (on the agency site and increasingly on central 'consultation gateways', for example the *ConsultWA* Catalogue: http://www.citizenscape.wa.gov.au/index.cfm?fuseaction=catalogue.about);
- conventional advertising through mass media; and
- selective recruitment of key stakeholders.

The tools enabled by ICTs can be useful in developing innovative and effective means of recruitment, particularly in difficult-to-access segments of the community.

4.5.1. Conventional Advertising and Promotional Approaches

Any eEngagement process, correctly configured to take account of issues associated with the digital divide, will incorporate a conventional promotional strategy (such as advertising, direct mail / marketing, etc.). Public sector managers will need to confirm their agency's policy (or wider government policy) governing the use of advertising (e.g. preferred vendor lists, timing issues, branding strategies, etc.).

In addition, the management team will also consider:

- the appropriate integration of ICT-based information with advertising (such as referral from advertisements to informative websites);
- the lead-in times for purchasing advertising (which can take months to schedule);
- careful project planning to ensure online materials are ready to go 'live' at the scheduled start time for the promotional campaign; and
- leveraging the 'novelty' of the process to ensure media coverage of the event (a valuable public relations approach that can deliver cost-effective coverage of the issue).

4.5.2. The Power of Social Networking (and its Limitations)

'Social networking' or 'referral' (or even 'multilevel') marketing is the use of existing social networks (such as friendship groups) to spread promotional and recruitment messages. These approaches are already used in consultative approaches in government, either formally ('bring a friend') or organically, as information relevant to one person is spread by them to their friends and personal acquaintances.

> **Exhibit 26: Wellington Shire (Victoria, Australia) Council Webcasting**
>
> The Shire of Wellington introduced web-based video cameras into Council meetings to extend the reach of Council meetings to its large shire. As part of the implementation the Council included the capacity for viewers to post questions to the Mayor (following formal meeting practices for gallery observers), which made the process more engaging and interactive, as the Mayor responded live on camera to public concerns and questions. The webcasts have had strong viewer numbers, partially because of good promotion and marketing via existing media channels including television coverage on the regional news and positive endorsement by the local newspaper. In addition, some journalists who live at a distance to the Council chamber use the system to cover council debate, increasing the 'knock on' effect of information distribution and oversight of Council activities.

Social networking is particularly powerful in the ICT world, where messages can be spread quickly and easily in digital form (as when colleagues forward messages about issues they think may be of relevance to persons in their social or professional networks). In addition, many websites include 'mail this page to a friend' options to allow people to easily distribute information they think of interest to people they know.

The advantages of including a social networking recruitment element can include:

- low/no cost;
- ease of implementation;
- the 'networking effect' can massively multiply the number of people who receive the message; and
- the message is targeted to people who are likely to be responsive.

The limitations of social networking include:

- uncertainty about the number of people who are likely to be recruited;
- 'sameness' of friendship groups (may need to seed many different groups to get a diversity of participants);
- risk of recruiting 'wrong' or ineligible participants (remember the global nature of the medium);
- loss of control over the communication as it passes along personal networks;
- loss of control over the timing of messages (particularly where social networks are infrequent communicators) – may lead to request for participation long after the eEngagement process has concluded; the difficultly in developing

effective messages (e.g. that have intrinsic appeal and are, therefore, likely to be passed along); and
- the need to ensure that recruitment is not undertaken in a way that would be seen as deceptive or in violation of relevant Privacy laws.

4.6. Managing Risk

Most, if not all, public sector managers are now familiar with the main tenets of risk management as a key process in project management. Many of the risk assessment and mitigation processes in current use are well-documented 'checklist' approaches. This means, however, that they sometimes suffer from 'over-formalisation'. The introduction of eEngagement processes will be undertaken with reference to potential risk.

James L Creighton[5] provides a useful checklist to assess the level of controversy associated with a topic, an important precursor to the development of appropriate risk management and mitigation strategies. According to Creighton, public managers need to ask:

- are the impacts of the change/issue significant?
- has there been prior controversy?
- does the issue tie into others that have a history of controversy?
- does the issue touch on politically charged issues?
- is this issue the *raison d'etre* for stakeholder groups?

By using this form of assessment tool, the level of potential controversy can be determined and particularly sensitive issues or groups identified. While Creighton observes that there is no 'mathematical formula' for the identification of levels of sensitivity, this type of risk assessment approach is something that (a) can assist in planning for the avoidance or minimisation of risks and (b) offer an important accountability mechanism if risks become manifest in the process.

This assessment may be developed simply as a mitigation process, however, where risks cannot be mitigated fully, the process will also be necessary as a means of providing information about risks to potential participants. This is particularly true with regard to privacy issues (as discussed in Section 3.3, Managing Identity Issues), where the capacity to provide a completely private environment for participation is limited, due to the agency's lack of control over the user's ICT platform (e.g. they may have an insecure personal computing environment). Beyond privacy, the most common issues of concern are security and defamation.

---

[5] Creighton, James L, 2005, *The Public Participation Handbook: Making Better Decisions Through Citizen Empowerment*, Jossey-Bass, San Francisco.

4.6.1. Security

Security is a technical *and* social concern and relates to:

- the safety of participants – particularly if their privacy cannot be guaranteed. While this may be irrelevant in many cases, experiences with consultation on issues of family violence in the UK necessitated careful planning and support to foster the participation of victims, particularly where they were alienated from their partners or remained in an at-risk environment; [6]
- the integrity of the system – even if the consultative process is not contentious, any networked system is open to attack and vandalism. Standard security procedures will be taken to prevent intrusion (which could lead to the loss of personal data of participants) or prevention of access attacks. [7] Where the issue is contentious, extra levels of security (higher security investment, distribution of hosting machines, multiple redundancy) should be applied to prevent disruption to the engagement process. This must be done in consultation with technical managers and security experts (often IT support staff may not have expertise in this area and external advice needs to be considered);
- social issues – it is important to recognise that most breaches of online security result either from 'insider' attacks (internal staff misuse of the system) or where users are 'tricked' into giving away identifying information ('social engineering'). Careful design of the consultation approach, appropriate management of staff with access to the system, (e.g. preventing access by staff to areas of the software or database not relevant to their work) and training for users (to resist social engineering attacks) can reduce these risks significantly. [8]

Exhibit 27: Open Source for Security

In the development of the eVACS electronic voting system, the ACT Electoral Commission released the source code of the software as open source. This release allowed third party organisations and individuals to identify and report problems with the code. See: http://www.elections.act.gov.au/EVACS.html

---

4.6.2. Moderation

Moderation (monitoring and exercising editorial control over message content) is necessary in some areas of eEngagement and has generated a number of

---

[6] Coleman and Gøtze's *Bowling Together*. The full reference is included in *Further Reading*.
[7] An attempt to prevent access to the system by 'flooding' it with fake users.
[8] An additional benefit here is increasing the public's awareness of online security issues more generally – a growing area of public policy concern.

practical reference guides and formal training (the Hansard Society in the United Kingdom runs an online course for moderators). Moderation can be necessary where:

- the issue is contentious and emotions can run high;
- the target audience is new to online communications (and may be naive about the implications of contributing to online conversations and their ability to be widely read);
- the contents of discussion is intended to be published (in whole or part);
- there are political or cultural sensitivities associated with online discussion or debate; and
- participants are young.

While these issues are not relevant to simple interactive approaches (such as one-off data collection, or the use of polling and surveys), the most commonly cited risks or concerns of public sector managers are:

- the presentation of material from participants (online or off) opens the agency to the risk of defamation (e.g. they are acting in the role of a *publisher*); and
- aggressive, lurid or rude postings to a discussion list can lower the tone of conversation – either reducing the tenor of conversation (generating little of value) or intimidating potential participants (silencing).

Both are real risks, with the latter more serious than the former.[9] The role of public officials (or third party moderators) in maintaining a correct tone of discussion is important – even if this is simply to keep debate and discussion 'on topic' and focused towards the consultative objectives.

The difficulty in determining an appropriate approach is often:

- failure to consider this issue before the project is initiated (thereby lacking rules and technical processes for moderation if problems emerge);
- lack of experience in many public sector organisations in moderation – the nature of online communications – its lack of paralinguistic cues and other 'social' indicators makes online moderation a specific skill set that needs to be developed and cultivated over time; and
- inappropriate setting of the level or extent, of moderation. Overly *light* moderation is as bad as having no moderation at all, while excessively draconian control of discussion (allowing no off-topic conversation at all, which undermines the 'forming and norming' social bonding process, for

---

[9] While defamation laws have been applied to the online environment, it has been recognised that publishers (sponsoring agencies or companies) only have limited control over the content posted, provided they act in good faith (a post hoc or complaints-based takedown approach). Additionally, the Australian online censorship laws (the *Broadcasting Services Amendment Act 1999*) provide exemptions for dynamic content that is not stored as a static resource (e.g. the correspondence of emails over an unachieved list are exempt from censorship under the Act).

example) can prevent the appropriate 'flow' of conversation that can make these processes largely self-directing (thereby reducing staff time in 'prompting' and 'guiding').

Comprehensive moderation can be expensive, requiring considerable allocation of staff to the task (depending on the number of participants). This is particularly true where moderation requires all communications to be read in real time (such as may be required in a chat room for young people, for example). However, a number of options exist to maintain a robust approach to moderation at lower cost. Which options the organisation employs will depend on the nature of the issue and participants, but can include:

- using a mix of paid staff and volunteer moderators;
- using keyword searching to identify suspect posts for human review;
- ensuring participants are not anonymous; and
- using a ranking system to allow readers to 'vote down' offensive or irrelevant posts.

The advantages and limitations of different approaches are outlined in Figure 11.

Figure 11: Advantages and Limitations of Moderation Approaches

Approach	Advantages	Limitations
Gate Keeping (pre-posting review and approval)	• Strong control over content, limits risk of hijacking or defamation • Focused discussion reduces 'off topic' conversation • Clear rules of engagement and participation, useful for inexperienced participants	• Overly controlling, can limit valuable tangential discussion • Can alienate participants • Limits 'discovery' function – data collection can be railroaded toward expected conclusions • Significantly slows conversation
Post-hoc Moderation	• Manages risks without excessive control • Guiding role of moderation can stimulate participation from shy participants • 'Referee' function can build community and reduce tension in complex and contested issues	• Time consuming, especially where high degrees of negotiation are required • Can still attract criticisms of control or censorship • Slows free-flow of discussion
Unmoderated (open forum)	• No risk of accusation of censorship • Low cost • Free flowing discussion • Can allow discussion to flow to unexpected areas (discovery)	• Risks of hijacking or defamation (modest) • Can lead to domination by small number of vocal contributors • Discussion can drift towards irrelevancy

Exhibit 28: Handling Defamation in a Discussion Forum

The City of Brisbane (Queensland) maintains a clear policy for managing issues of defamation on its citizen discussion lists (http://ycys.brisbane.qld.gov.au/). This policy consists of:

- a formal policy statement that is provided to participants when they subscribe to the service;
- a moderation process that sees a member of the city council staff review all messages before they are posted to the list; and
- a complaints handling process with avenues for appeal and review.

The aim of this process is to protect the City from publishing material which may result in an action for defamation, or lead to general incivility on the discussion list. Items that are deemed to violate the policy are:

- in the first instance referred back to the original author pointing out the areas of difficulty and with suggestions as to how the message may be modified to comply with the policy; and
- subject to review (upon request) by a more senior manager for final determination.

e-democracy.org, on the other hand, maintains a 'take down' approach, where messages that violate the rules of the discussion (http://www.e-democracy.org/rules/) are removed if they are deemed to violate the rules.

5. Concluding the Process

One of the most significant issues in developing an effective eEngagement process is careful planning of the post-implementation activities for the project. This has two elements:

- developing an appropriate and robust approach to *meaningful* evaluation, particularly when there is a need to justify the activity in a highly rational (budgetary-focused) operational environment – an increasingly common concern; and
- developing an effective *closeout* process.

5.1. The Importance of Evaluation

There is little need to reiterate the importance of evaluation in the public sector. Calls for discussion of debate around and methodological experimentation with evaluation have been hallmarks of public sector management reforms for the past decade. Any project initiated in the public sector today will make provision for evaluation as a standard operating procedure.

In the context of a new type of activity, however, careful consideration of evaluation is important. This is because:

- while most (if not all) governments in Australasia stress the importance of public participation and engagement, the practical commitment of governments is often quite variable. The relative newness of these activities often creates an environment in which novel or innovative approaches to community engagement are often subject to higher levels of scrutiny and assessment. This, combined with the potential to generate greater levels of feedback about the process itself, can put the innovating public sector manager under a degree of scrutiny not shared by managers following 'tried and true' (but possibly ineffective) strategies to engage the public;
- the area is new and requires grounded, honest evaluation of the cost and benefits of a range of different approaches. While it is likely that eEngagement will continue to be an important part of the armoury of public sector managers for the foreseeable future (if not increasingly important over time as our society develops greater levels of technical sophistication and complexity), effective and practical evaluation of the vast array of models and techniques will lead to better means to assess the benefits of one approach over the other, making planning faster, implementation easier and the outcomes more effective; and
- the use of ICTs can support new approaches to evaluation, increasing the effectiveness of this part of the management process and leading to higher levels of understanding about what works and what does not, than in offline

activities. This is a direct result of the interactivity of the media employed and their capacity to support the automatic collection of user data.

5.1.1. Approaching Evaluation for eEngagement

The exact nature of evaluation will be highly variable depending on the mechanisms and approaches employed (and objectives). Whyte and Macintosh[1] provide a useful conceptual tool for evaluating eEngagement activities, focusing on political, technical and social outcomes of the project or process. This approach is recommended for any eEngagement activity and asks the following questions:

- *political evaluation:* Did the process follow best practice guidelines for undertaking consultations that are published by government and were the stakeholders satisfied with the process? The evaluation factors here are similar to those for conventional consultations but need to be answered by different means;
- *technical evaluation:* To what extent did ICT design directly affect the outcomes? In designing the e-consultation there is a need to take account of the technical skills of the target audience and locality of the participants. Here, we can take as our starting point established evaluation frameworks from the software engineering and information systems communities, together with considerations of usability and accessibility; and
- *social evaluation:* To what extent did the social practices and capabilities of those being consulted affect the consultation outcomes? In particular, what bearing do these have on the relevance of consultations to the consulted citizens, the relevance of their contributions to each other and to policy makers and the nature of the interaction?

5.1.2. Pitfalls to Avoid

Common traps to avoid in developing the evaluation framework are:

- *over-emphasis on technical assessments:* Technical issues are often easy to document and can be clearly presented in terms of equipment 'up time', budgetary management and ease of systems implementation. While these issues are important, it is important to keep them in perspective and not lose sight of the broader objectives (e.g. technology merely facilitates the process, it is not the end product);
- *excessive use of simple metrics:* Many consultation processes are assessed purely on the basis of number of participants, or amount of content generated. While this has an important role, it is critical to also ask:
    - 'right people' versus 'many people';

---

[1] Whyte, Angus and Macintosh, Ann 2003, 'Analysis and Evaluation of E-Consultations', *e-Service Journal*, vol. 2, no. 1, <http://www.e-sj.org/e-SJ2.1/esj2_1_whyte_macintosh.pdf>

- what are the characteristics of the people engaged (e.g. were 'new' people brought into the process, does information flow though these people to a wider audience – are they 'influentials'?); and
- *picking the right comparisons:* If the eEngagement process has been implemented to assess a consultation or participation deficit, the particular approach used will be assessed against the previous state of play (before-after assessment), rather than with other examples that use the same technology or methodology – these latter types of comparisons are often of limited value.

5.1.3. What to Consider in Effective Assessment

When developing the assessment approach, it is important to consider:

- the extent to which the technology can support longitudinal assessment processes (e.g. performance measurement over time, or reducing a long 'end of process' survey into a series of small polls);
- user views and experiences (sometimes best expressed qualitatively). Consider allowing the users to develop and present their own evaluation frameworks (a variation on self-assessment reporting);
- 'knock-on', capacity-building, or social capital formation outcomes (skills transfer, mobilisation, organisational outcomes and benefits); and
- the development of real-time and automatic metrics. A good example of this would be the ability to incorporate comprehensive analyses of user browsing patterns with respect to online information (e.g. pages viewed, time spent viewing each page, pages with highest levels of referral to others, etc.). These metrics allow us to analyse (for better or worse) the value of our content in a way that print run numbers of consultation documents cannot. These statistics can often be provided by the service provider (such as the website hosting service or from the telecommunications provider) or through commercial services (e.g. Nielsen//Netratings).

> **Exhibit 29: Evaluation Example – Local Issues Forum Success Measures (longitudinal)**
>
> | | |
> |---|---|
> | 2-3 Months | • forum is still active
• some regular traffic
• experiencing some membership growth
• city / community officials are aware of forum / may be reading posts
• some community organisations have begun to post announcements in forum |
> | 6 Months | • 25-50 percent growth in subscriptions since launch
• local media is to paying attention to discussions
• 10 or more 'regular' posters (post at least once per week)
• participants are starting new discussions
• regular participation in steering committee communications and meetings attract a diverse group of community members |
> | 1 Year | • elected officials and city / community staff are participating – most lurk, but some post
• 50-100 percent growth in subscriptions since launch
• occasional story in local media that originates from forum
• some examples of citizen or government action that have resulted from forum discussions
• you have hosted at least one in-person gathering or party for participants to meet one another |
>
> E-Democracy.org 2005, *Local Issues Forum Guidebook*

5.2. Closeout Processes

A common failing of many consultation processes is a failure to consider and plan for the end of the eEngagement process. This tends to reflect an instrumental view of the process which holds that, once the information has been collected or the decisions reached, the engagement is over.

This can lead to:

- a failure to fully and appropriately document the process when the lessons of the process are freshest; and
- 'orphaning' the participants, either by not providing them with appropriate levels of information about the outcomes, or by neglecting a possible valuable future resource of interest to stakeholders.

Clear planning for the closeout process will require:

- an appropriate commitment of time (staff time);
- a publication schedule for information (feedback); and
- possible re-investment in *cultivation* of the stakeholder community.

5.2.1. Document Process and Outcomes

It goes without saying that the relative newness of eEngagement, combined with the rapid pace of change (both in the capacities of the technology and the costs of undertaking activities using ICTs) mean that – for the immediate future at least – practice will continue to outstrip theory.

Following the conclusion of any engagement activity – online or off – it is necessary to prepare a suite of post-engagement documentation which normally takes the form of:

- formal reports on outcomes;
- internal reports on project management (costs, user responses, etc.);
- public seminars and debriefings; and
- 'bottom line' accountabilities (budget reporting).

Given the newness of this area of activity, it is important for many of these (often internal) documents to be shared with the eEngagement community, i.e. those who are actively pursuing the area of practice, those interesting in undertaking activities and those not aware of the potential. This often necessitates the development of case study information – the repackaging of information provided to a range of stakeholders in a complete encapsulated form.

Good case documentation will include:

- a clear outline of the background (issue, agency, jurisdiction, culture);
- an articulation of what type of initial decisions were made;
- discussions of technologies employed;
- a discussion of activities, including unforseen issues;
- evaluations of outcome (short, medium and longer term and a 'balance of assessment' statement);
- unresolved issues;
- issues for future application – often these processes generate large numbers of innovative ideas that cannot be taken up at the time, but would be of great value to managers contemplating emulating the model; and
- contact information (including for partner organisations).

One of the important aspects of this documentation needs to be a clear statement of the managerial *learnings*: namely, the 'lessons learned' at the managerial level about handling 'intangibles' (such as upwards and downward management, stakeholder issues, etc.). While there is an excellent array of case studies now being developed, attention to subtle management questions will be one area of particular interest to others in your position.

5.2.2. Feedback

A common criticism heard from many consultation and engagement participants is the lack of feedback from government agencies on the outcomes and decisions made from the information received.

Maintaining good post-engagement relationships is important in maintaining citizens' motivation for civic participation and the inclusion of eEngagement projects can be a valuable means by which feedback is delivered at low cost. The low cost of email, fax and SMS communications, together with their capacity to deliver multimedia content, makes the provision of feedback relatively straightforward and can stimulate further, or future, participation from members of the target community.

Feedback should contain:

- a summation or means by which large documents/information can be accessed quickly and easily (such as an appropriate executive summary for a formal reporting process, fact sheet, or information bulletin);
- notification of the results of the eEngagement process: what decisions have been reached, what plans or processes are to be implemented, where the issue has advanced in the decision-making process (if the consultation is an early part of a longer process of policy development);
- where there is significant variation of opinion or disagreement, balanced reporting of the range of opinions or perspectives and information about reasons for the selection of specific options (either because of majority voting outcomes in deliberative processes, or the basis for decisions made in purely consultative ones);
- collection of 'opt in' permissions to contact the participant again for future engagement processes (either on a similar or unrelated subject) to develop a larger database of stakeholders; and
- 'big picture' views about the scope of the eEngagement process, such as the number of participants, timescale, etc. Where eEngagement processes involve little or no personal interaction, participants can often lack a sense of the number of other participants (unlike in the traditional 'town hall' style meeting) and so knowing the scope of participation will place the legitimacy of the outcome in context.

The provision of feedback regarding specific instrumental (policy specific) outcomes of the process can be an appropriate point in which stakeholder views on the conduct of the engagement process can be collected (if this has not already been done). It is important to note that the quality and nature of feedback provided at the closeout stage of the process will also be assessed for future reference.

> **Exhibit 30: Maintaining Contact – Address Lifecycles**
>
> When collecting contact information from participants (to allow information to be 'pushed' to them), the limited 'life expectancy' of contact information must be considered. While email is often considered an excellent communication channel because of cost and speed, it can also be highly *temporal*.
>
> Consider the limitations of various channels based on the life expectancy of their use:
>
> - email addresses are notoriously short lived, possibly lasting only between 1-3 years on average. This is often associated with changes to ISP connections, employment changes and the tendency to 'shed' addresses that have become targets for high volumes of SPAM messages. People who have a lasting valid email address tend to be in long-term employment. The life expectancy of *Instant Messaging* addresses (such as Microsoft or Yahoo! Messenger, Skype, etc.) is unknown at this time, but may also be short;
> - residential addresses are relatively long lived, approximately 7-8 years on average, however, this average is highly variable and tends to be a function of stage-of-life (marriage, children) and the age of the individual. As a general rule, the younger the adult, the more likely they are to change residential address; and
> - mobile telephone numbers may prove to be one of the most enduring contact addresses for participants in eEngagement processes, particularly following the introduction of MNP (mobile number portability – the capacity to retain a fixed mobile telephone number even following changes of service). Australia introduced MNP in 2001 and New Zealand is expected to do so in 2007-8.
>
> Given the short life expectancy of contact addresses and telephone numbers, it is wise to collect a number of contact details from participants for future engagement and follow-up. Delivery failure using one channel can then prompt the use of alternative approaches.

5.2.3. Feedback Over Time

In some policy deliberations, it may be wise to establish an ongoing process of feedback provision to participants. This helps maintain public interest in the issue and personal commitment to participation by citizens.

On-going feedback is most appropriate where:

- the policy development process is ongoing (e.g. the eEngagement process was at the start of a wider policy-development process, such as a

parliamentary consultation, where executive decision-making supersedes the initial eEngagement activity);
- the consultation leads to a policy or project implementation process, allowing participants to observe the translation of policy into public action;
- a subsequent executive decision has reversed or significantly altered the initial conclusions drawn from the consultation process (a change of policy); and/or
- there is a desire to stimulate an active, or passive body of concerned citizen oversight, such as the establishment of a transparency network. In this case, allowing citizens access to a shared community space where they can contact each other and discuss the issue will be required. This type of approach can serve as the interface between formal eEngagement processes and wider eDemocracy stimulation and capacity building.

Exhibit 31: Transparency Networks

Transparency Network is a term used to describe organically connected groups of organisations and individuals who share information and oversight of the activities of policy makers, government agencies and corporations. The participants of the network can include policy insiders, non-government organisations, scholars, journalists and members of the community. Based around loose network organisational structures and using ICTs, (email, discussion lists, websites), these networks collect and distribute information and can act to highlight issues or problems that emerge in their area of concern. Good examples of transparency networks can be found in the environmental movement, where large numbers of quasi-autonomous actors and groups can mobilise and organise over environmental issues and policy processes.

By nature, these networks are outside of government and largely outside of formal eEngagement processes (though members of transparency networks are often found in formal consultation and participation processes). Governments are increasingly responsive to these networks, both positively (providing greater access to oversight information, inclusion in consultation processes) and negatively (secrecy), depending on the ability of the networks to utilise their members' resources to challenge policy decisions and implementation (often in tandem with mainstream media). Transparency networks share many similarities with the notion of 'policy communities' from mainstream public policy literature, but may take a more 'outsider' role.

See: http://www.agimo.gov.au/publications/2004/05/egovt_challenges/accountability/transparency

5.2.4. No Closeout: The Eternal Community

While careful management of the closeout process can involve ongoing communication with participants, the conclusion of a formal process of eEngagement may not mean the 'end' of the process. For example, instrumental processes often lead to the creation of on-going communities of interest or relationships with the hosting agency through the development of formal reference groups, participants transition from 'passive' to active overseers of government policy and the future re-use of consultation mailing lists.

In addition, in areas where the expected benefits of the engagement strategy are broad and diffuse, the project may have an expectation of stimulating the development of a 'community of interest' around the policy area or agency that is relatively self-sustaining over time. Clearly, the toolsets provided by ICTs to the public to self-organise and network outside the direct intervention of government, represent a key strength.

eEngagement processes can result in the mobilisation of an ongoing community of interest. Public sector managers can be instrumental in fostering these communities of interest via a cultivating approach and drawing value from them by exercising a listening role. Public sector managers should consider the following:

- has the process generated support for the creation of an ongoing community of interest?
- do the participants have the tools necessary to act on their desire to maintain an ongoing relationship *with each other?*
- what benefits would this provide to ongoing policy development and implementation (and hence, what is the cost-benefit of stimulating activity)?

Examples of active roles public sector managers may play to cultivate these types of ongoing outcome are:

- ensuring information flows to participants;
- planning a listening strategy after the closeout of the formal eEngagement process;
- cultivating interactions between stakeholders through the provision of toolsets (email list software, wiki engines, etc.) to the community;
- 'rewarding' communal activity through *ad hoc* or informal meetings or gatherings; and
- determining means by which 'listening to the community' can be demonstrated (e.g. periodic email contributions to discussion lists on topics raised in these communities, pro-actively taking forward issues of concern, etc.).

The end may be just the beginning of a new phase of engagement.

Exhibit 32: Wiki's and Collaborative Tools

A 'wiki' is a popular term for collaborative software which allows anyone participating in the development of the content to edit what is published or presented. Good examples of wiki's include the free online general encyclopaedia Wikipedia (http://en.wikipedia.org/wiki/Main_Page) or the Davis Community Wiki (http://daviswiki.org/).

Wiki's require the establishment of motivated communities, authoring and collaboration tools, storage space and mediating and arbitrating processes for managing version control. Other examples of collaborative approaches to online publishing would include:

- slashdot (http://slashdot.org/)
- e-the People (http://www.e-thepeople.org/)

Further Reading

- Bimber, Bruce. 2003. *Information and American Democracy: Technology in the Evolution of Political Power* . Cambridge University Press, Cambridge.
- Boak, Cathy and Blackburn, Jean. 1998. *So, You Want to Host an Online Conference…* Human Resources Development Canada, Ottawa, <http://www.hrsdc.gc.ca/en/hip/lld/olt/Skills_Development/OLTResearch/ConfGdNode_e.pdf>
- Clift, Steven. 2002. *The Future of E-Democracy – The 50 Year Plan* . <http://www.publicus.net/articles/future.html>
- Coleman, Stephen and Gøtze, John. 2001. *Bowling Together: Online Public Engagement in Policy Deliberation*. Hansard Society, London, <http://bowlingtogether.net/>
- Davis, Richard. 1999. *The Web of Politics: The Internet's Impact on the American Political System*. Oxford University Press, Oxford.
- E-Democracy.org. 2005. *Local Issues Forum Guidebook* . E-Democracy.org, <http://e-democracy.org/uk/guide.pdf>
- European Commission. 2004. *Seminar Report – eDemocracy* . eGovernment Unit, Information Society Directorate General, European Commission, Brussels, <http://europa.eu.int/information_society/activities/egovernment_research/doc/edemocracy_report.pdf>
- Green, Lyndsay. 1998. *Playing Croquet with Flamingos: A Guide to Moderating Online Conferences*. Human Resource Development Canada, Ottawa, <http://www.hrsdc.gc.ca/en/hip/lld/olt/Skills_Development/OLTResearch/flamingo_e.pdf>
- Hansard Society. 2005. *Members Only?–Parliament in the Public Eye* . Hansard Society, London, <http://www.hansardsociety.org.uk/programmes/puttnam_commission/launch>
- Harvard Policy Group. 2002. *Eight Imperatives for Leaders in a Networked World: Imperative 8 – Prepare for Digital Democracy* . Harvard Policy Group on Network-Enabled Services and Government, John F. Kennedy School of Government, Cambridge, <http://www-1.ibm.com/industries/government/ieg/pdf/Imp8.pdf>
- Johnson, David, Headey, Bruce and Jensen, Ben. 2003. *Communities, Social Capital and Public Policy: Literature Review*. Melbourne Institute Working Paper No. 26/03, Melbourne Institute of Applied Economic and Social Research, The University of Melbourne, <http://melbourneinstitute.com/wp/wp2003n26.pdf>
- Local Government Commission. 2003. *Geographic Information Systems: A Tool For Improving Community Livability*. Local Government Commission,

- Sacramento, <http://www.lgc.org/freepub/PDF/Land_Use/fact_sheets/gis.pdf>
- Lukensmeyer, Carolyn and Torres, Lars. 2006. *Public Deliberation: A Manager's Guide to Citizen Engagement*. IBM Centre for the Business of Government, <http://www.businessofgovernment.org/main/publications/grant_reports/details/index.asp?GID=239>
- Norris, Pippa. 2001. *Digital Divide: Civic Engagement, Information Poverty and the Internet Worldwide*. Cambridge University Press.
- OECD. 2003. *Engaging Citizens Online for Better Policy-making*. Public Affairs Division, Public Affairs and Communications Directorate, OECD, Paris, <http://www.oecd.org/dataoecd/62/23/2501856.pdf>
- OECD. 1998. *Impact of the Emerging Information Society on the Policy Development Process and Democratic Quality*. Public Management Service Public Management Committee, OECD, Paris, <http://www.olis.oecd.org/olis/1998doc.nsf/c16431e1b3f24c0ac12569fa005d1d99/a968ff0e858cf2a48025675400628e3b/$FILE/12E81094.DOC>
- Queensland Government. 2005. *E-democracy Evaluation Framework*. Queensland e-Democracy Unit, Department of Communities, Brisbane, <http://www.getinvolved.qld.gov.au/share_your_knowledge/documents/word/eval_framework_summaryfinal_200506.doc>
- Queensland Government. 2004. *Engaging Queenslanders: A Guide to Community Engagement Methods and Techniques*. Department of Communities, Brisbane, <http://www.e-democracy.gov.uk/documents/retrieve.asp?pk_document=182&pagepath=http://www.e-democracy.gov.uk:80/knowledgepool/>

Online Resources

- Access2democracy: http://www.access2democracy.org/a2d/content/en/index.aspx
- Australia:
  - AGIMO E-democracy Community of Practice (Australia, Commonwealth): http://www.agimo.gov.au/resources/cop/e-democracy
  - Citizenscape – Western Australia: http://www.citizenscape.wa.gov.au/
  - Nation Forum eDemocracy Resource Site: http://democracy.nationalforum.com.au/
  - Victorian Electronic Democracy Inquiry Report: http://www.parliament.vic.gov.au/sarc/E-Democracy/Final_Report/ToC.htm
  - Queensland e-Democracy Unit: http://www.communities.qld.gov.au/community/edemocracy.html
- Canada:
  - Canadian Online Consultation Technologies Centre of Excellence: http://www.pwgsc.gc.ca/onlineconsultation/text/index-e.html

- Crossing Boundaries Working Group on Democratic Reform and Renewal: http://www.crossingboundaries.ca/democratic-renewal-en.html?page=democratic-renewal&lang_id=1&page_id=155
  - Electronic Commons: A Public Network: http://www.ecommons.net/
- Commonwealth Centre for Electronic Governance: http://www.electronicgov.net/
- Deliberative Democracy Consortium: http://www.deliberative-democracy.net/
- European Forum for European e-Public Services: http://www.eu-forum.org/
- IBM Institute for Electronic Government: http://www-1.ibm.com/industries/government/ieg/
- International Institute for Democracy and Electoral Assistance: http://www.idea.int/
- New Zealand:
  - eDemocracy.co.nz: http://www.edemocracy.co.nz/
  - eGovernment in New Zealand: http://www.e.govt.nz/
  - Good Practice Participate: http://www.goodpracticeparticipate.govt.nz/
  - Naturespace: http://www.naturespace.co.nz/ed/
  - Waitakere eDemocracy Group: http://www.wedg.org.nz/resources.html
- OECD:
  - e-Government Project: http://webdomino1.oecd.org/COMNET/PUM/egovproweb.nsf
  - Directorate for Science, Technology and Industry: http://www.oecd.org/department/0,2688,en_2649_33703_1_1_1_1_1,00.html
- United Kingdom:
  - Active Citizenship Centre: http://www.active-citizen.org.uk/
  - Oxford Internet Institute: http://www.oii.ox.ac.uk/
  - Society of Information Technology Management E-Government Exchange: http://www.socitm.gov.uk/exchange
  - UK Local e-Democracy National Project: http://www.e-democracy.gov.uk/
- United States:
  - America Speaks: http://www.americaspeaks.org/
  - Congress Online Project: http://www.congressonlineproject.org/
  - Digital Government Research Centre: http://www.diggov.org/

Appendix A. Policy Cycle Engagement Model

Stage in policy-making cycle	Information	Consultation	Participation
Agenda-setting	• site-specific search engines • email alerts for new issues • translation support • style checkers to remove jargon	• online surveys and polls • discussion forums • monitoring emails • bulletin boards • frequently asked questions	• e-communities • e-petitions • e-referenda
Analysis	• translation support for ethnic languages • style checkers to remove jargon	• evidence-managed facilities • expert profiling	• electronic citizen juries • e-communities
Formulation	• advanced style checking to help interpret technical and legal terms	• discussion forums • online citizen juries • e-community tools	• e-petitions • e-referenda amending legislation
Implementation	• natural language style checkers • email newsletters	• discussion forums • online citizen juries • e-community tools	• email distribution lists for target groups
Monitoring	• online feedback • online publication of annual reports	• online surveys and polls • discussion forums • monitoring emails • bulletin boards • frequently asked questions	• e-petitions • e-referenda

Source: 'Box 2: Tools for online engagement at each stage of policy-making', *Policy Brief: Engaging Citizens Online for Better Policy-making*, OECD Observer March
2003, © OECD 2003.

Appendix B. Catalogue of eEngagement Models

An alphabetically-arranged list of general models of eEngagement.

Contestable Policy Analysis

Interactivity:	Low
Timeliness:	Variable
Outcomes:	Variable
Decision-making:	Government
Complexity:	Low
Description:	The notion of contestable policy analysis is a broad one and tends to be less programmatic, or project based, than other forms of eEngagement (such as consultative models). This notion comes from broader policy discussions regarding contestability in government service delivery and is an extension of new public management concepts of competition across all facets for government. Contestable policy analysis is a deliberate attempt to ensure that aspects of policy analysis are open for participation by non-government organisations and individuals, be they private sector firms (through contracting), academic organisations, or partisan groups. The essential requirement for contestable policy analysis is to ensure that information relevant to the assessment of policy options is released to the public. In the past, these have generally taken the form of comprehensive policy discussion papers, where public sector understanding of the issue and research is summarised for public consideration. However, with the advent of ICTs, the cost of delivering greater amounts of information and the capacity for external groups and individuals to analysis large amounts of data, has increased significantly. Thus, while discussion papers generally included statistical evidence, it seldom included complete data sets. The analysis of policy in a contestable manner requires external actors access to the same amount and form of data as internal analysts.
Advantages:	<ul><li>transparency and oversight by members of the public (can introduce new evaluations of data quality or assumptions employed by the public sector that are not known to executive and elected officials)</li><li>stimulation of competing perspectives and analysis (can be of higher quality than internal analysis)</li><li>generation of alternative policy ideas, based on sound analysis</li><li>non-directive – capacity for new ideas to be generated</li><li>cheap – digital release of information is very low cost</li></ul>
Limitations:	<ul><li>loss of control over data. If data has value (is employed commercially) care must be taken to establish an appropriate licensing system (such as the Creative Commons approach, see: http://creativecommons.org/) to preserve public ownership but allow contestability</li><li>issues of privacy. Some data may identify (or be able to be used to infer) individuals. The assessment, cleaning, or aggregation of these data sets can incur a cost</li><li>no guarantee of external expertise, or that expertise will be applied (uncertain outcomes)</li><li>political sensitivity to the release of data (as opposed to the release of carefully prepared documents, as is commonly the norm) can be at odds with trends towards greater government control over information release and presentation</li><li>concept misuse – selective data release ('good news data' only) will lead to distorted analysis (garbage out, garbage in)</li></ul>

Co-production (eGovernance)[1]

Interactivity:	High
Timeliness:	Ongoing
Outcomes:	Specific
Decision-making:	Shared
Complexity:	High
Description:	Co-production models of engagement focus on shared policy-making and management between government and the community or relevant non-government organisations. Whereas partnership and outsourcing models tend to focus on relationships that are either based on principal-agent models (such as contractual relationships) or devolution and autonomy (self-government), co-production entails equal participation by both parties and recognises this through shared decision-making functions. The use of ICTs in this area can include: • developing online joint management boards or structures that use ICTs to overcome distance issues, or problems of participation out-of-hours (either through virtual meetings or the provision of briefing and performance data electronically to reduce time commitments in physical meetings) • integration of management systems (such as performance and reporting systems) across organisations, where policy implementation is a joint undertaking (data sharing and aggregation) • creation of 'virtual organisations' with staff and budgets allocations drawn from a range of organisations (public value creation networks) • development of performance data exchange systems between purchaser-provider organisations (vertically), across separate geographic delivery areas (horizontally), or to form performance markets (comparative and competitive environments to determine true and contingent cost per performance evaluations)
Advantages:	• co-production is aimed at achieving inter-organisational collaboration and action. Thus, can represent an expression of 'joined-up' government. The use of ICTs can allow existing structures (hierarchical bureaucratic departments) to be 'overlaid' with co-production networks to achieve this, without radical restructuring • capacity to achieve better policy outcomes (e.g. matching resources with expertise, expertise with local implementation) – magnification effect • focuses on information exchange, joint development of programmatic responses and shared management (true partnerships) • democratic and participative – can overcome significant barriers to implementation • coalition building in character (can overcome entrenched interest problem)
Limitations:	• accountability issues • complexity (especially in data exchange systems development) and therefore cost implications • need to establish flexible systems to accommodate change can widen scope of initial network development at a cost (e.g. need to develop extensible data exchange and collaboration protocols and applications) • often difficult to achieve where large differences in resources exist between partners (elephant and mouse problem) • need to recognise value of non-economic (financial) resources to develop meaningful partnership models • 'drift' between outcomes of self-managing networks and top-down (executive) policy making can be problematic

[1] Often referred to within a 'partnership' framework. However, it is asserted here that the misuse of the term partnership – particularly for public financing arrangements, which represent a form of monopoly licensing; in areas of Australian indigenous governance and; as a misnomer for consultation – has undermined the value of this term in an engagement context.

Online Citizen Juries

Interactivity:	High
Timeliness:	Short to modest
Outcomes:	Specific
Decision-making:	Public
Complexity:	Modest to High
Description:	Citizen juries are small groups of citizens (normally 10 to 15 members) who are brought together to hear evidence related to a policy issue, deliberate amongst themselves and pass a resolution. The approach differs significantly from a focus group, in that the length of time undertaken is longer as the jury is presented with evidence from experts on the subject prior to their deliberations. The outcome of the citizens jury is either a binding resolution or a recommendation which, if not implemented, must be responded to. The use of ICTs to facilitate this form of decision-making approach can allow for participation asynchronously (expanding the number of people who can participate who would normally be restricted by work or carer commitments), present evidence from a wider range of experts who may be based internationally and provide evidence in a range of forms (multimedia, written)
Advantages:	<ul><li>as a form of direct decision-making this approach has been heralded as having democratic value, in that decisions are seen to be taken by 'ordinary' people</li><li>appropriate use of sampling for jury selection can gather a broad cross-section of the community, or reflect a specific community composition which may be distinctly different to that of the public sector or elected representatives</li><li>the ability to provide expert evidence and place this within the public arena can improve overall understanding of the complexity of decision-making</li><li>can often be a useful approach to break through a policy area where decision-making has been dominated by an entrenched interest</li></ul>
Limitations:	<ul><li>can be expensive and sometimes cynically utilised to provide a veneer of legitimacy</li><li>the selection of experts can highly shape outcomes</li><li>the small number of participants can be used to question the legitimacy of the outcome, particularly where the decision reached is widely divergent from popular opinion</li><li>failure to implement jury decisions can breed disenchantment and scepticism over the honesty of the commitment to engagement</li><li>often unsuited to highly technical areas of policy-making</li></ul>

Online Deliberative Conferencing

Interactivity:	High
Timeliness:	Modest
Outcomes:	Specific
Decision-making:	Public
Complexity:	High
Description:	A variation of online citizens' juries, deliberative conferencing dramatically increases the scope and scale of the undertaking and can include as many as several thousand participants divided into small groups that come together for plenary sessions and to hear evidence. Online deliberative conferencing draws its claim to strength from the large number of participants and the capacity to sample a broad cross-section of the community. The large number of participants does require meticulous planning and a significant investment in the systems that allow the views of each of the small groups to be incorporated into a final share outcome. This is often undertaken through the use of a series of surveys or polls undertaken throughout the course of the event. • a non-deliberative form of this approach is sometimes referred to as 'community visioning'. In this type of approach final outcomes are often highly qualitative, rather than passing specific resolutions or endorsing particular final conference policy documents
Advantages:	• largely identical to the online jury, with the advantage of broader participation • useful in large-scale 'visioning' exercises and can be useful in developing significant public commitments to a substantial change of direction in public policy • the large investment of time required to develop materials serves as a significant resource in educating the public about a policy issue
Limitations:	• extremely expensive, particularly in the development of technology, the recruitment of participants and the recruitment and training of staff. Often volunteers are employed to lower these costs • like citizen juries, questions are often raised about the practical utility of undertaking these activities. While deliberative conferences have shown that the opinion of a broad cross-section of the community can be shifted given a rigorous and complete briefing and discussion of highly charged policy issues, this does not mean that the wider community will endorse these views having been largely outside of the process • again, like citizen juries, the selection of people to provide expert evidence can be highly contested and attract the accusation of manipulation through bias in the selection of these experts

Electronic Delegate Committees

Interactivity:	Modest to high
Timeliness:	Ongoing
Outcomes:	Specific
Decision-making:	Shared
Complexity:	Modest
Description:	Electronic delegate committees have similarities to citizen juries in that they are comprised of small numbers of citizens who have some claim to represent a segment of the community. In this model, this claim is based on the election from specific groups, rather than the 'market research' sampling approach of the citizen jury or deliberative conference. Delegate committees meet to discuss policy issues, exchange information about the perspectives of their respective groups or communities and can have a specific deliberative all decision-making function (devolution of decision-making).
Advantages:	<ul><li>by using elected delegates from specific target communities all organisations can overcome some of the criticisms of the sampling approach utilised in citizen juries and deliberative conferencing, as delegates are directly elected</li><li>delegates can act as an information conduit between their community or representative body, multiplying the information transfer effect at modest cost</li><li>appropriate briefing or training for delegates can assist in improving the quality of deliberation, provided delegates caucus or survey their constituents</li><li>effective use of elections can provide significant legitimacy to outcomes</li></ul>
Limitations:	<ul><li>difficulties in establishing specific communities or bodies to be represented can lead to problems in establishing an effective electoral system, undermining legitimacy of process</li><li>elected nature of delegates can create tension with conventional political processes where delegates claim political legitimacy above formal elected representatives</li><li>recruitment process can be time-consuming and expensive to establish and administer to prevent electoral fraud</li></ul>

Electronic Discussion Lists

Interactivity:	High
Timeliness:	Ongoing
Outcomes:	Specific or diffused
Decision-making:	Variable (may contain voting engine for direct or deliberative decision-making)
Complexity:	Low to modest
Description:	Often based on relatively simple technical systems (such as bulletin board systems or email list servers) they are relatively simple to develop. Depending on the purpose of the discussion list the process can be discreet (e.g. subject specific) or ongoing, canvassing a wide range of topics for discussion (the 'reference group' model). Electronic discussion lists can be strictly controlled through moderation or limits to the number of contributions from participants in a given period of time, or can be open and unregulated. Some electronic discussion lists have employed 'chat' software (such as Internet Relay Chat) to host real-time discussions. • a variation of this approach is the use of these technologies to undertake online focus groups (closed lists) as an asynchronous substitute for conventional face-to-face approaches
Advantages:	• flexible format can be empowering to participants, allowing members of the public to define the subject under discussion and engaging conversation between themselves • relatively simple technology employed can be useful as a low barrier to entry and participation (particularly where email is the delivery channel) • collect large amounts of data, with a high degree of interactivity allowing for follow-up discussion over unclear aspects of the conversation • can be used as part of an online citizens jury model, with the use of voting or polling at key points
Limitations:	• often requires a considerable time commitment from participants, which can be a barrier to participation • sometimes difficult to recruit participants, particularly where the issue is considered dry • where moderation is not undertaken, can result in domination by a small number of participants or external disruption, such as flooding with 'spam' messages • can be over moderated, restricting discussion, often due to a fear of 'hijacking' by partisans • data collected can be highly unstructured and, without specific voting mechanisms, analysis can be difficult • moderation can be expensive

Electronic Voting

Interactivity:	Low
Timeliness:	Short (periodic)
Outcomes:	Specific, quantifiable
Reach:	Broad
Decision-making:	Public
Complexity:	Very High
Description:	Electronic voting systems have been introduced in a number of countries, particularly the United States, Canada, United Kingdom and India, with mixed success. The primary motivation for the introduction of these systems is generally as a means to combat declining levels of participation in noncompulsory electoral systems and a range of technologies has been employed, from standalone or locally networked personal computers, mobile telephones, internet-based systems, to specifically built voting devices. Additional benefits attributed to the introduction of these systems emphasise their capacity to deliver verbal instructions in a variety of languages and allow voting remotely
Advantages:	<ul><li>builds upon existing participatory paradigms well understood by means of the public</li><li>fast tabulation of election outcomes, particularly in complex electoral systems</li><li>multilingual and vision impaired assistance</li><li>remote participation</li></ul>
Limitations:	<ul><li>highly expensive</li><li>high risk environment, particularly for internet-based voting systems</li><li>low public trust in technical system</li><li>limited public demand in Australasia</li></ul>

Online Dispute Resolution

Interactivity:	High
Timeliness:	Short (issue specific)
Outcomes:	Specific, focused on resolving disputation
Decision-making:	Public
Complexity:	Medium to High
Description:	Online dispute resolution is an emerging area of practice that stems from conflict resolution studies and has attracted strong interest from some aspects of the legal and judicial community. While it is commonly employed to resolve personal or commercial disputes (and thus is an ideal complement to electronic commerce), this approach can be employed to resolve local area disputations in an environment that can be divorced from the intensity of face-to-face interaction. This can be particularly valuable where one or both of the parties feels intimidated.
Advantages:	<ul><li>has proven to be an effective way of resolving disputation, particularly where parties have had a breakdown of relationships which makes face-to-face interaction counterproductive</li><li>asynchronous nature of communication affords benefits where parties are in different timeframes, or have incompatible working commitments (e.g. where a citizen-based group is in conflict with a commercial organisation)</li><li>can employ decision support technology to model minimum agreements conditions</li><li>electronic nature of communication can be used to document agreements reached during activity</li></ul>
Limitations:	<ul><li>relatively new and emerging area of practice, may face resistance from entrenched stakeholders</li><li>can move political disputation into a closed arena and away from public scrutiny</li><li>potential resistance from existing dispute resolution professionals</li></ul>

Electronic Surveys and Polling

Interactivity:	Low
Timeliness:	Short, but can be used on an ongoing basis as part of a reference group
Outcomes:	Specific, but can be used as a precursor for less structured consultation and participation processes
Decision-making:	Government
Complexity:	Low
Description:	The use of ICTs to deliver surveys to the community has been well developed over the last decade and the proliferation of low-cost, easy-to-use online publication tools makes the development and implementation of these engagement processes relatively simple to deliver
Advantages:	<ul><li>low-cost to develop and deliver, elimination of data entry costs and transcription errors</li><li>ease of delivery and completion can increase response rates</li><li>'smart' surveys can include error checking and dynamic presentation of complex surveys that reduce the prevalence of accidental submission of incomplete surveys</li><li>easy importation of collected data into analysis packages</li><li>easy collection of contact information for follow-up research</li></ul>
Limitations:	<ul><li>generalised difficulties determining identity</li><li>impersonal nature of approach</li><li>can be overly rigid</li><li>tendency to emphasise quantitative results over qualitative ones (particularly in large samples)</li><li>'polling' approaches sometimes lead to trivialisation through over focusing on either/or questions where the issue is complex</li></ul>

Simulations and Games

Interactivity:	High
Timeliness:	Short
Outcomes:	Specific
Decision-making:	Government
Complexity:	High to Very High
Description:	The use of planning simulations and other types of policy-oriented games have had application in public consultative processes for over 50 years. Often, the intention of these approaches is educative: either pitched towards younger citizens (such as children as part of civics education) or adults. The advantages of these approaches are their high level of interactivity, engaging nature and capacity to illustrate a range of policy alternatives (good simulations have far more combinations of policy response than their designers could envisage). With the development of ICTs and their popular use in gaming, these games can be: • highly complex and sophisticated – giving good insights into the trade-off outcomes of a range of policy alternatives (high-end simulations) • graphically impressive • delivered remotely • allow for large-scale competitive or collaborate play (network gaming) • delivered across a range of platforms (stand alone programs, web-interfaces, interactive television, etc.) Simulations and games can be developed at a range of levels, from corporate-grade decision support simulation, to modifications of existing game engines used for popular play,[a] to simpler implementations based on text or web-based animation tools (such as Flash).
Advantages:	• compelling, engaging content • can make participation 'fun' • educative – can show immediate and long term impact and outcomes of policy decisions (projections) to participants (for example, the Australian Stock Exchange share market simulation game: http://www.asx.com.au/investor/education/games/index.htm) • can allow for solo and network (collaborative play) • simulation decisions can be stored and 'submitted' as preferred plans (e.g. a simulation allowing citizens to develop an optimal arrangement for inner city land use can allow different preference maps to be stored, published and voted upon) • high levels of information literacy in these online environments, particularly amongst the sub 35-year old age group • once established, can be 'self managing', can create communities of interest around the game
Limitations:	• often high costs of design, often requires long lead time • issue needs to be well understood for appropriate and accurate simulation development (a *contestable* simulation makes the underpinning assumptions of the model clear and able to be changed by end users to see outcomes under alternative interpretations of the issue or problem) • rarely used outside of planning purposes – may be difficult to secure stake holder commitment • can be seen as trivialisation of a serious issue (see, for example, the United Nations 'Food-Force' game: http://www.food-force.com/) • real-time games are unsuitable for people with low levels of experience in gaming, or who have limited dexterity

[a] The United States armed forces, for example, licensed the popular game engine for Quake to develop a recruitment-oriented combat game (http://www.americasarmy.com/). In addition, there are a range of open source simulation engines (http://sourceforge.net/softwaremap/trove_list.php?form_cat=85).

www.ingramcontent.com/pod-product-compliance
Lightning Source LLC
Chambersburg PA
CBHW060947170426
43197CB00031B/2989